smart simple design

/ reloaded /

Happy Thanksgiving —
Steve,

Gwendolyn

smart simple design

/ reloaded /

Variety Effectiveness
And The Cost Of Complexity

Gwendolyn D. Galsworth

Visual-Lean® Enterprise Press
An Imprint of Visual Thinking Inc.
Portland, Oregon, USA

©2015 Gwendolyn D. Galsworth. All rights reserved under International and Pan-American Copyright conventions, including the right of reproduction in whole or in part, in any form.

Visual-Lean® Enterprise Press
Division of Visual Thinking Inc.
607 NE 32nd Avenue
Portland, Oregon 97232
503-233-1784 (phone)
info@visualworkplace.com
www.visualworkplace.com

Galsworth, Gwendolyn D.
ISBN: 978-1-932516319

Editor: Aurelia Navarro
Book and Cover Design: Iwan Sujono at eOne Design

Visual-Lean®, Visual Office®, and Visual Machine® are federally registered service marks, globally licensed to Visual Thinking Inc.

Visual Thinking Inc. (VTI) and the author hope you study, appreciate and learn from this book and apply the principles and practices of variety effectiveness in your own company. Please be aware that this book is a work of original authorship and the materials in it are protected by US and International copyright law. This book is not in the public domain. Therefore, if you or your company want to acquire training materials so you can teach, educate, and support others in *Smart Simple Design/Reloaded* or other of Dr. Galsworth's many improvement technologies, contact us directly. Further, you must make written arrangements with us before excerpting from this book or developing materials based on it. It is illegal to reproduce this publication in any form or for any reason whatsoever, or store it in a retrieval system, or transmit it in whole or in part in any form or by any means whatsoever, including electronic, mechanical, photocopying, recording, or otherwise without the prior written permission of VTI.

Disclaimer. This publication is designed to provide accurate and respected information in regards to the subject matter covered. It is sold with the understanding that neither the publisher nor the author is engaged in rendering professional legal, engineering, technical or accounting services. If advice or other expert assistance is required, the services of a competent professional person in the associated field should be sought. Both the publisher and author, therefore, specifically disclaim any liability, loss or risk, corporate, organizational or personal or otherwise, which may occur as a consequence, directly or indirectly, of the reading, use or application of any part of the contents of this book, which book is sold with that understanding in mind.

Dedication

To Geraldine Galsworth, mother and source of inspiration, strength, and joy—thirty-nine years young has never looked more beautiful on anyone.

And to the Unity that is God, Who makes all parts—however diverse and however various—One.

> Making the simple complicated is commonplace. Making the complicated simple—awesomely simple—that's creative. **CHARLIE MINGUS**

By The Same Author

Books

- Visual Workplace-Visual Thinking
- Work That Makes Sense: Operator-Led Visuality
- Visual Systems: Harnessing the Power of the Visual Workplace
- Visual Workplace-Visual Order Associate Handbook
- Visual Workplace-Visual Order Instructor Guide
- Smart Simple Design (1994)

Training and Implementation Courses

- Work That Makes Sense: Operator-Led Visuality
- The Visual Machine®
- Quick Changeover and The Visual Machine®
- Achieving Perfect Quality Through Mistake-Proofing
  (co-author, Dr. Martin Hinckley)
- Visual ScoreBoarding: Problem Solving for the Chronic, Complex, and Costly
- Becoming a Leader of Improvement: The Supervisory Function
- Becoming a Leader of Improvement: The Executive Function
- Management by Sight: Visual Displays/Visual Scheduling
- Smart Simple Design/Reloaded: Variety Effectiveness and the Cost of Complexity

DVD Training System

- Visual Workplace-Visual Order *(with Spanish subtitles)*

Smart Simple Design/Reloaded
Contents

List of Figures and Illustrations	xiii
Foreword by Richard J. Schonberger	xviii
Foreword by Bruce E. Hamilton	xix
Acknowledgements	xxiii

Part I — The Cost Of Complexity

Introduction	3
Chapter 1 The Consuming Age: Relentless Pressures of a Voracious Marketplace	7
The Age Of Disinflation Has Arrived	9
The Challenge	10
The Iceberg Effect: Hidden Danger	11
Variety Effectiveness	12
About This Book: Chapter By Chapter	14
Chapter 2 On the Horns of the Dilemma	17
PUI: Exponential Growth/Less Profit	17
Visual Evidence of a Problem	18
What Went Wrong?	20
The Improvement Revolution	21
The Push for New Products	25
Expanding Choices—Collapsing Cycles	27

Exploding Variety: One New Part	28
Too Much of Too Much	29
The Eight Runaway By-Products of One New Part	31
Rethinking the 95:5 Ratio—Death by 10,000 Cuts	33
New England Farmhouse Effect	35
PUI: When Variety is Negative	36
VEP: The Alternative	37
The Rewards	38
What is Variety Effectiveness	40
X-Type Company	41
VEP's Multi-Dimensional Approach To Effective Variety	42
1. Comprehensive View of Cause	43
2. In-Depth Analysis of Complexity	44
3. Insight-Rich, Team-Based Approach to Improvement	45
4. Unraveling Complexity through Engineering Based Tools	45
5. Prevention-Based Transformation	46
VEP: Not Magic—Work!	46

Chapter 3 True Cost: Product Proliferation & the Bottom Line — 49

A Case In Point: Accounting Gone Awry	50
The Traditional Cost Approach: History and Logic Of GAAP	52
From Tracking the Past to Tracking Complexity	54
Flaws In Traditional Cost Accounting Assumptions	56
Flawed Assumptions About Cost	56
Flawed Assumptions About Price	57
The Allocation Of True Cost: A New Cost Perspective	59
The Origins of Organizational Complexity	59
Cost Adhere to Parts: The Part as First Cause	60
Measuring Product and Organizational Complexity	62
The VEP Parts Index: Universal Measure of Complexity	63
Not an Exact Measure—But an Exacting One	68
The Index as an Improvement Driver	69
The VEP Tri-Cost Model: The Three Dimensions Of True Cost	69
Dimension 1: Functions Costs	70
Dimension 2: Variety Costs	72
Dimension 3: Control Costs	73
The VEP Cost Pie	74
Control Points and ABC Accounting	75
Awareness is Everything	76

Chapter 4 Negative Variety and Its Policy Triggers 77

Variety Explosion: Unintended Consequences 78
   Negative vs. Positive Variety 79
   The Balance Point: Effective Variety 81
The Policy Roots Of Negative Variety 81
   A. Negative Triggers in Accounting Policies 84
      1. Supplier Base Selection 84
      2. Make vs. Buy Decisions (Cost per Part) 85
      3. Overhead Cost Allocation 86
   B. Negative Triggers in Marketing and Sales Policies 87
      4. Response to Customer Requests 88
      5. Margins, Product Pricing, and Discounting 89
      6. New Product Market Requirements 90
      7. Cost Targeting/Cost Reductions 91
   C. Negative Triggers in Product Development Policies 92
      8. Product Diversification Approach 92
      9. Products as Separate Entities 95
      10. Long Lapses between Products 95
      11. Different Designers/Different Design Concepts 95
      12. Cost Targeting in New Product Development 96
      13. Technical Solutions 96
      14. Product Life-Cycle Decisions 97
   D. Negative Triggers in IT/Data Systems Policies 99
      15. Product Documentation, Classification Systems, and Computer Support 99
   E. Negative Triggers in Operations Policies 100
      16. Lot Sizing 100
      17. Capital Equipment and Process Improvement Justification 101
Stem The Tide 102

Chapter 5 Hot Products: Design for Overall Cost 111

CAD: A Double-Edged Sword 112
   Over-Designing: Overusing Your Strengths 112
Revising The Mind-Set: Design For Overall Cost 113
   Design from the Outside In—Put Value Near the User 114
   Know When Average is Good Enough 116
   Use Fewer Parts—Use Shared Parts 116
   Get Sales Involved—From the Get-Go 117

Barriers To Moving Forward	118
Six Barriers to Effective Variety	120
A New Role To Play	122

Part II VEP: The Methodology

Chapter 6 The VEP Methodology
Stage 1: Getting Ready to Launch — 127

Deciding On Scope	127
The Discrete Approach	128
The Deep Dive Approach	129
Overview Of The VEP Method	129
Details Of Stage 1: Prepare For An Effective Implementation	132
VEP Teams	133
VEP Leadership Teams	134
VEP Analysis Teams	135
VEP Support Teams	138
Select Your Starting Point: The Targeted Series	144
The Qualifying Procedure	144
Next In VEP's Stage 1	148

Chapter 7 Creating a VEP-Capable Classification System — 149

Venerable Chair Company	149
Juanita Hicks to the Rescue	150
Taking On The Job	151
A Case In Point: A Bulging, Bungling, Blundering Data System	152
Where Group Technology Fits In	158
Making The Data System VEP-Capable	159
Begin with Nomenclature	159
Second: Define Attribute Templates	161
Third: Tackle the Class Codes	163

Chapter 8 The Six VAT Tools — 165

The Six VATS: Tools Of Inquiry	165
VAT-1: Unique vs. Shared	166
The Benefits of Sharing	168

VAT-2: Modularity	170
Three Modular Styles	172
V-Costs and the First Two VATs	174
VAT-3: Multi-Functionality & Synthesis	174
VAT-4: Ease Of Assembly	177
VAT-5: Range	179
Other Applications of Range	182
VAT-6: Trend	183
The Power of the Six VATS	185

Chapter 9 The 3-View Analysis — 187

Why Three Views?	187
Do We Really Need this Difference?	189
View One: Market Analysis	190
Your Nomenclature: The Groundwork	191
PUI's Product Hierarchy	191
Calculating the Combination Magnitude	193
Comparing Product Attributes	194
Implications: Market Analysis Process	197
Spreading Out the Reduction Net	197
View Two: Product Structure Analysis	198
Key Elements of the Product Analysis Procedure	198
Implications: Product Structure Analysis	203
View Three: Parts Type Analysis	203
Parts Type Analysis at PUI	206
Implications: Parts Type Analysis	209
Systemic Problem—Systematic Approach	212

Chapter 10 Reducing Downstream Complexity — 213

Negative Variety In Processes	215
Reducing Processes at PUI	215
First: Standardize the Nomenclature	216
Second: Reduce Processes	219
Negative Variety In Control Points	220
The Search for Control Points at PUI	222
The Control Points Reduction Team and Its Tasks	223
Downstream Is Where The Silt Piles Up	227

Chapter 11 Implementing VEP Improvements — 229
  Stage 1 Review: Prepare — 229
  Stage 2 Review: Analyze By Applying The Six Vats — 231
  Stages 3 And 4 — 231
    Stage 3: Prioritize and Schedule — 231
    Stage 4: Implement and Prevent — 234
  Making The Change: Some High-Level Implementation Issues — 236

Part III The Bottom Line

Chapter 12 Designing for the Bottom Line — 243
  Variety Effectiveness: A Unified Approach — 247
  The Goal Is To Make More Profit — 249

Resources — 251

Glossary — 253

Visual Thinking Inc. & The Visual-Lean® Institute Resource Page — 261

Index — 263

About the Author — 271

Smart Simple Design/Reloaded
List of Figures

CHAPTER 1

Figure 1.1.   Danger! The Real Problem Remains Hidden   11

CHAPTER 2

Figure 2.1.   The Seven Deadly Wastes + One   22
(Non-Value-Adding Activity)
Figure 2.2.   Rocks in the River: The Flow of Production   23
Figure 2.3.   Value-Adding Activity vs. Non-Value-Adding Activity   24
Figure 2.4.   The Eight Runaway By-Products of New Product Expansion   32
Figure 2.5.   The Seven Deadly Wastes of Production: 33
Equivalents in New Product Development
Figure 2.6.   95:5 Ratio in Runaway By-Products Hidden in New   34
Product Activity
Figure 2.7.   Blocks in the Production Flow: Rocks Revisited   34
Figure 2.8.   New England Farmhouse Effect to New Product Introduction   35
Figure 2.9.   The Y-Type Trajectory: Profile of a Company in Trouble   36
Figure 2.10.   VEP Outcomes: A Chain of Rewards   40
Figure 2.11.   The X-Type Curve: Profile of a Company Succeeding   41
Figure 2.12.   The Five Power Points of the VEP Approach   43

CHAPTER 3

Figure 3.1.   True Total Cost: Which Product Costs More? (Comparison   52
of Series 11 and 8 Products)
Figure 3.2.   GAAP vs. VEP   55
Figure 3.3.   Consequences of Adding Just One New Product with   61
Just One New Part
Figure 3.4.   Parts List: Model J-191/Blue Pen   62

Figure 3.5.	Parts List: Model J-192/Red Pen	63
Figure 3.6.	VEP Parts Index for Models J-191/Blue and J-192/Red	64
Figure 3.7.	VEP Parts Index for Models J-191, 192, 193, 194	65
Figure 3.8.	VEP Parts Index: Partial BOM for 3 of 51 PUI Models	66
Figure 3.9.	Three Dimensions of True Cost Defined	71
Figure 3.10.	The VEP Cost Pie	74

CHAPTER 4

Figure 4.1.	Effective Variety: An Ever-Shifting Balance Point	82
Figure 4.2.	Negative Variety: 17 Policy Triggers	83
Figure 4.3.	Improved Policies for Trigger 1: Supplier Base Selection	85
Figure 4.4.	Improved Policies for Trigger 2: Make vs. Buy Decisions	86
Figure 4.5.	Improved Policies for Trigger 3: Overhead Cost Allocation	87
Figure 4.6.	Improved Policies for Trigger 4: Response to Customer Requests	88
Figure 4.7.	Improved Policies for Trigger 5: Margins, Product Pricing, and Discounting	90
Figure 4.8.	Improved Policies for Trigger 6: New Product Market Requirements	91
Figure 4.9.	Improved Policies for Trigger 7: Cost Targeting and Cost Reductions	92
Figure 4.10.	Improved Policies for Triggers 8 to 14: Product Development	98
Figure 4.11.	Improved Policies for Trigger 15: IT and Data Systems	100
Figure 4.12.	Improved Policies for Trigger 16: Lot Sizing	101
Figure 4.13.	Improved Policies for Trigger 17: Purchase Justification	102
Figure 4.14.	Summary: All Policy Triggers of Negative Variety	103

CHAPTER 6

Figure 6.1.	VEP Methodology: Stage 1	131
Figure 6.2.	Memo: Company Commitment to VEP	132
Figure 6.3.	VEP Teams for a Deep-Dive Implementation	133
Figure 6.4.	The Chuck Wagon: Early Victories Team	139

Figure 6.5.	Sample: VEP Team Configuration, with Time Commitments	141
Figure 6.6.	Modified Relations Diagram	146
Figure 6.7.	PUI's Top Five Series: Score Sheet for Pre-set Qualifying Criteria	147

CHAPTER 7

Figure 7.1.	Before VEP: PUI Active Class Codes (104 Total Codes)	154
Figure 7.2.	Before VEP: Report of Active Screw Part Numbers (Partial List)	156
Figure 7.3.	Before VEP: Examples of PUI's 30-Character Codes	157
Figure 7.4.	After VEP: Active Class Codes at PUI (Reduced from 104 to 74 Class Codes)	159
Figure 7.5.	After VEP: PUI's Attribute Template for Active Screws	161
Figure 7.6.	Class Code Guidelines for Parts	163

CHAPTER 8

Figure 8.1.	Unique vs. Shared Parts Index: Series Level	169
Figure 8.2.	Summary of VAT-1: Unique vs. Shared	170
Figure 8.3.	VAT-1 and VAT-2: Two Perspectives on the Same Parts	172
Figure 8.4.	Summary of VAT-2: Modularity	173
Figure 8.5.	VAT-3: Multi-Functionality & Synthesis - Bracket (Before/After)	175
Figure 8.6.	VAT-3: Multi-Functionality & Synthesis - Plunger (Before/After)	176
Figure 8.7.	Summary of VAT-3: Multi-Functionality & Synthesis	176
Figure 8.8.	VAT-4: Ease of Assembly - Spring (Self-Aligning and Self-Locating Parts)	177
Figure 8.9.	VAT-4: Ease of Assembly - Pin (Parts Cannot Be Installed Incorrectly)	178
Figure 8.10.	VAT-4: Ease of Assembly - Housing (Adequate Access/Unrestricted Vision)	178
Figure 8.11.	Summary of VAT-4: Ease of Assembly	179
Figure 8.12.	VAT-5: Range - O.D. Values in a Scatter Diagram	180
Figure 8.13.	VAT-5: Range - O.D. Values in a Histogram	181
Figure 8.14.	Summary of VAT-5: Range	182
Figure 8.15.	VAT-6: Trend in Requests for Miniaturization (Six-Month Intervals)	184
Figure 8.16	Summary of VAT-6: Trend	184

CHAPTER 9

| Figure 9.1. | VEP Methodology: Stage 2/The 3-View Analysis | 188 |
| Figure 9.2. | PUI's Top 20 | 192 |

Figure 9.3.	PUI's Product Hierarchy: Nomenclature of Levels	192
Figure 9.4.	Example: Combination Magnitude on Three of PUI's Top 20 Product Series	194
Figure 9.5.	Market Attribute Matrix (Partial): Series Level	195
Figure 9.6.	PUI Market Analysis Team: Reduction Recommendations (First Pass)	196
Figure 9.7.	Visual Layout of BOM	199
Figure 9.8.	Partial BOM: One of the 02 Models	200
Figure 9.9.	VEP Parts Index: Partial BOM for 02 Models in Series 7, 8, 11, 33, and 97	201
Figure 9.10.	Model 02 Reduction Recommendations (Partial List)	202
Figure 9.11.	Sample: VEP Parts Profile Work Sheet	205
Figure 9.12.	Part Types in Series 7 (Partial List)	206
Figure 9.13.	VAT-5: Range - Springs Re-Visited by the Parts Type Team	207
Figure 9.14.	Parts Type Reduction Recommendations: Screws and Brackets	208
Figure 9.15.	VEP Team Meeting Form: Market Analysis	210

CHAPTER 10

Figure 10.1.	VEP Methodology: Stage 2/ Reduction Analysis for Processes and Control Points	214
Figure 10.2.	Before VEP: Names for PUI Production Processes	217
Figure 10.3.	After VEP: Names for PUI Production Processes (Grouped by Category)	218
Figure 10.4.	PUI Process Attribute Template: Welding	219
Figure 10.5.	A Part Sprouting Control Points	220
Figure 10.6.	Control Points: What Really Happens	221
Figure 10.7.	Control Points: Associated with XJ-889 Sub-Assembly	222
Figure 10.8.	VEP Control Points Index: Accounting Forms	225
Figure 10.9.	Sample: VEP Improvement Work Sheet for Forms	226
Figure 10.10.	Actual Control Point Reduction Results (Partial)	227

CHAPTER 11

Figure 11.1.	VEP Methodology: Stages 3 and 4	230
Figure 11.2.	Sample: VEP Proposal Impact Worksheet	233
Figure 11.3.	Sample: 15-month Implementation Timeline	237

Foreword
by **Richard J. Schonberger (2014)**

Gwendolyn Galsworth is already well established as the world's premier advocate of visual systems and all that goes with it. Dr. Galsworth should be equally well known for *Smart Simple Design/Reloaded*. The book's subject matter penetrates the core of manufacturing excellence because getting product design right generates rippling benefits throughout the business—all the way to the customer. The book's subtitle sums it up elegantly: *Variety Effectiveness and the Cost of Complexity*.

Galsworth doesn't use the usual narrow term, variety *reduction*, which could raise hackles, especially in sales and marketing. Instead she chooses variety *effectiveness*, meaning product variety that reduces *total cost* while maintaining optimal *customer selection*.

Smart Simple Design/Reloaded broadens the usual technical approaches to intelligent design by creating an impressive superstructure—the organization of the simple-design effort. Galsworth calls it the Variety Effectiveness Process (VEP), offering keys to a successful, enduring implementation. Far too many manufacturers deal with effective design only in fits and starts—clearly due to the lack of such an over-arching methodology. A seminal chapter, "Getting Ready to Launch," gives thorough guidelines on where and how to begin, and an easy-to-follow chart that summarizes the entire process.

VEP does much more than organize the effort. Within VEP are tool-like devices eminently usable by design engineers in collaboration with others involved in the process. Prominent among them is the *VEP Parts Index*, a no-nonsense way of counting parts and part-types interactions that add up to "a universal measure of complexity." Complexity, of course, creates compounding sets of costs. Chapter Three is a fine look at "True Cost: Product Proliferation and the Bottom Line." In it, Galsworth blends cost issues with a convincing case study example of growing numbers of parts in a product's bill of materials (BOM). The example includes showing how each additional part in a BOM multiplies complexity and cost effects.

In this brief foreword I'll just mention two of the other highlights in the book. In Chapter Four, Galsworth addressed the Seventeen Policy Triggers of Negative Variety, characterized as the "variety explosion." These seventeen triggers are distributed across five departmental functions and then enumerated, one by one: Accounting, Marketing and Sales, Product Development, IT/Data Systems, and Operations. Then Galsworth turns each of these negative triggers on its head with her recommendations for new and improved policies that support VEP's transformative outcomes.

Also worth a strong mention are the *Eight Runaway By-Products of One New Part*, discussed in Chapter Two. As the words suggest, parts and product proliferation—however modest—nearly always come with many undesirable companions or by-products and those proliferate too. This chapter also makes a good case for looking at these as equivalent to the classic Toyota waste wheel but on a whole new level.

I like this book—a lot. Part of the reason, I suppose, is that its emphasis on simplicity fits so well with a central theme of my own first book, *Japanese Manufacturing Techniques (1982), the subtitle of which was Nine Hidden Lessons in Simplicity.*

It makes me wonder if, at the time, I had had the luck to have Gwendolyn as co-author, whether the subtitle might have been something like *Twelve Hidden Lessons in Simplicity.* Such a book would have been filled out with strong chapters on variety effectiveness, raising it to stature as one of the primary methodologies of JIT production and lean manufacturing.

Still, today, among books numbering in the hundreds on lean and related topics, it is hard to find erudite treatments of product proliferation and variety effectiveness. Dr. Galsworth's book is the rare, fine exception.

You'll like this book, too.

Richard J. Schonberger, Ph.D.

Author (and originator of the term) *World Class Manufacturing: Lessons of Simplicity Applied* (Free Press, 1986), and a number of other books, including: Let's Fix It! (Free Press, 2001); *Best Practices in Lean Six Sigma Process Improvement* (Wiley, 2008).

Foreword
by **Bruce E. Hamilton (1994)**

When United Electric (UE) first began using the inventions of the Toyota Production System, we focused narrowly on our factory, seeking to eliminate waste from the production process. We were, as Shigeo Shingo put it, "constructively dissatisfied" with our production capability and determined to find a better way to manufacture. For the next three years, our success in reducing inventory and improving service was so great that it consumed our attention. As we examined and understood the huge waste from overproduction, for example, that had been created as we filled our stockrooms with large lots of partially completed product, we reduced lots from "nice round numbers" to the minimum order quantity. Then, as quick changeover techniques were employed, lot sizes were further reduced. Assemblies, previously produced in lots of 1,000 and then sorted for later use, were now triggered for production by a system that dictated that we build only what was needed, when it was needed, and in the smallest quantity determined by setup times.

All improvement could be measured by time saved: A reduction of the total elapsed time to fill a customer's order, a reduction in product development time to qualify a new supplier. Every invention we employed was directed to that end: Make the process go faster by eliminating waste. Three years later, our stockrooms had been eliminated and inventory had been reduced by 65 percent—many millions of dollars. Lead time dropped from months to days and perceived service was at an all-time high. Cellular production reduced flow distance from miles to feet and created a wholly new set of team and problem-solving skills required for production to work in a new way. For those of us in production, this was an exciting time in our careers. We were changing the way things were done, and we were changing ourselves in the process.

Just when it seemed things couldn't get any better, they didn't get any better. Concepts such as kanban, cellular manufacturing, single-minute exchange of dies, and poka-yoke had produced huge early results, but now were considered the norm, the basis for daily production. And on that basis, improvement leveled off. Inventory dropped so far, and then stopped dropping; likewise with lead times. While the cycle of improvement is never-ending, the tools used in that improvement cycle tend over time to become maintenance tools rather than improvement tools. What further tools could we identify to break through the improvement plateau?

Once we had stripped away some of the grossest forms of waste in our business—large inventories, useless material handling and storage equipment, even excess buildings—we began to see a major new opportunity for improvement. There was a huge cost, both in time and money, for every part in our system that was separate from the functional cost of the part. In production, we had learned how to identify waste in seven forms as taught by Toyota: storage, transportation, overproduction, unnecessary processing, motion, defects, and waiting. A year later, UE began working with Dr. Galsworth to establish a systematic method for identifying and eliminating an "eighth" waste—unwarranted variety.

Through the use of SMED (single-minute-exchange-of-dies), we reduced many lot sizes to one—but even for that one piece, we had to activate our entire production system. Now there appeared to be a means for reducing the variety costs associated with many parts by simply eliminating the part. This, in fact, we had addressed in a piecemeal fashion from the early days of our improvement process. However, in the absence of a clear method for measuring and identifying the trade-offs associated with variety—and more important, for understanding the root causes of variety—we seemed to be adding new variety at a greater rate than we were removing old variety. With thousands of parts and processes in our production system, the complexity of the problem dictated a new method for solution. This has evolved today into what Dr. Galsworth calls *variety effectiveness*.

Dr. Galsworth's book is the first thorough treatment of a method that can systematically identify the waste of needless variety. For older businesses especially, this book provides a blueprint for cutting back the mass of parts and processes resulting from years of product proliferation. But every business, established or start-up, can benefit from the methodology that extends the power of the Toyota Production System beyond production and into the design and development process. Most variety in part design, product structure, and process selection does not result from a customer's need, but from a series of internal policies and behaviors that needlessly complicate the production process. These include cost-accounting systems that actually reward part and process proliferation, engineering mores that eschew the use of previously developed designs, variety resulting from technology change, and inadequate design tools that actually make it easier for designers to develop a part from scratch than to search for an already-existing part. The customer is not in the equation.

The key benefit to designers in the method developed by Dr. Galsworth is that it supports broad selection for the customer while reducing variety in the product design in a manner transparent to the customer. By understanding which variety is negative, engineers and designers can contribute to profitability and service in a way that was not previously possible. VEP's simple but powerful techniques enable engineers to change the way they work, to work faster, and to develop products of exceptional selection that meet individual customer needs without adding layers of complication and cost to the production process. Dr. Galsworth's detailed process for improvement arms designers with a systematic method for identifying, classifying, and reducing unneeded variety.

For product marketers, the Variety Effectiveness Process represents an alternative to the all-too-common process of product-line trimming and selection retrenchment. By minimizing the cost and time of new product development, VEP brings more new products to customers sooner. And it revitalizes old products through its dramatic cost-reduction potential.

For production, the benefits are reduced part and process complexity, reduced equipment expense, reduced training expense, reduced material handling, improved turns, less stockouts, and fewer defects.

And for corporate management, there is the powerful message: Corporate structure mirrors product structure. Simplify the first and the latter will follow. For United Electric, VEP has offered the opportunity for a second wave of improvement. I view variety effectiveness as an approaching revolution in the product development process. Its effects will touch all aspects of an organization's competitiveness and financial wellbeing. Dr. Galsworth's book is an excellent place to begin this process.

Bruce E. Hamilton

(1994) Vice President/Operations
United Electric Controls Company
Watertown, Massachusetts

(2014) Executive Director
Greater Boston Manufacturing Partnership
Boston, Massachusetts

Smart Simple Design/Reloaded

Acknowledgements

This is the second time I have written *Smart Simple Design*. The first time was 25 years ago when I was engaged in developing the principles and practices for successfully implementing variety effectiveness. I was stunned by what I was discovering and anxious to put what I found into a form that would allow others to learn about it too. A book followed.

Published in 1994, that book defined and described the Variety Effectiveness Process (VEP) methodology I had developed for simplifying an enterprise by de-complicating its product architecture. Twenty years later, when I made the simple decision to upgrade and update that book, it took me nearly two years to complete the work.

The result is the book you now hold–*Smart Simple Design/Reloaded*—not just an update but a rigorously re-written volume whose main purpose is to ensure that the conceptual base of the paradigm is understandable, clear, and coherent. In the process, I re-aligned the book's arc, added several chapters, expanded its scope to include non-manufacturing environments, and clarified the role of de-complication in relation to two other powerful improvement strategies, visual and lean. I believe *SSD/Reloaded* is worthy of your time to read and digest, even if you are already well-acquainted with its predecessor. It contains many new elements as well as a fresh layout design and splendid new cover—both thanks to Iwan Sujono, an enormously gifted artist and book designer in Sydney, Australia.

While *SSD/Reloaded* is a substantially new book, when I turn to recognize those who helped me with this more recent task, I must first re-acknowledge those who helped me two decades ago—most especially:

Dave Reis, still president of United Electric Controls Company (UE) in Watertown, Massachusetts for his willingness to let me model aspects of the book's prototype "struggling" company (Parts Unlimited Inc.) after UE, an organization of exceptional achievement.

Bruce Hamilton, then VP of Operations at UE, for his recognition—in face of ballooning parts count—that old answers would not work, for collaborating with me on the VEP methodology, for his contribution to the policy chapter, for his conceptualization of the Early Victories Team, and for coining of the term "variety effectiveness"—vastly more appealing to sales and marketing than the earlier "variety reduction."

Patricia Wardwell, Maureen Hamilton, Bonnie Rafuse, Cheryl O'Connell, Bob Rando, and Barbara Murphy are just some of the many, many UE employees whose contributions to the VEP method were significant.

Toshio Suzue and Akira Kohdate for writing their seminal work, *Variety Reduction Program: A Production Strategy for Product Diversification* (Productivity Press, 1988 Cambridge, MA), with its fresh perspective on parts proliferation. Geoffrey Boothroyd and Peter Dewhurst of Boothroyd Dewhurst; William Hyde and Barry Levine of Brisch, Brin & Associates; and Norman Bodek, founder of Productivity Inc. and Productivity Press.

Patricia Moody, gifted author in her own right and editor of the 1994 book, and John Clegg, who contributed the majority of the 1994 illustrations, used again in *SSD/Reloaded*.

SSD/Reloaded is in great part the result of the vision, skill, patience, and professionalism of my lifelong friend and editor, Aurelia Navarro. She has been chief editing collaborator for nearly three decades across all eight of my published books and uncountable imprints, presentations, articles, and papers. Her unerring sense of a well-crafted sentence, tightly-arced chapter, and read-worthy books has never failed to serve. As with all the books that come under her regard, it is no exaggeration that *SSD/Reloaded* could not have happened without her.

I also send my heartfelt thanks to those others who over the past two years helped me as I undertook *SSD/Reloaded*—most especially:

Richard Schonberger, author of dozens of groundbreaking books on operational excellence, for his early and unflagging support of the VEP model—and for writing a fine Foreword to this volume.

Eric Lail, brilliant VEP practitioner and vice president at Transportation Insight, for his constancy and encouragement.

Alex Bleier, IT consultant and friend, for the application of his considerable technical skills to many aspects of this book. Suzanne Ralls, whose engineering background and MBA proved an invaluable assistance to many chapters. Martin Hinckley, nuclear engineer and friend, for his help on trending concepts. Manny "Ed" Velosa, member of UE's original 3-View Analysis Team, for his help in decoding ancient class code abbreviations for *SSD/Reloaded*. Daniel Forest and Roxane Vezina of Venmar Ventilation of Canada.

Professor Sir Michael Gregory, CBE, head of the Institute for Manufacturing at Cambridge University, whose dedicated sponsorship of additive manufacturing is changing the face of complexity—and its costs.

Michael Philpott, of Value Driven Design in the UK, for his on-going commitment to VEP deployment. And Jon Tudor, of True North Excellence in the UK, for his support of all of my work and many of my dreams.

My esteemed, high-spirited colleagues at Visual Thinking Inc.— Cindy Lyndin, Horatio Fairburn, Patty De'ak, Heidi Houston, Kelly McNiece, and Harald Hope: You make so much of my work have greater meaning.

My brother, Gary Galsworth, who simply loves me and wants to see my dreams come true—nothing more is needed because, in a brother, that is everything. And my nieces and nephew, Ondine Galsworth, Stacy Joyce, Karen Cathcart, and Daniel Galsworth, whose lives are shining reminders of the beauty of expression.

Every writer knows that dealing with the content portion of a book is only half the story. The other half is keeping one's spirit whole and body functioning. For their extraordinary help in this, I am deeply grateful to Mataare, Carolyn Hawkins, Barbara Paster, Dawn Bothie, Erick Wander, Clark Shea, and Merlin, my Cat.

And to Swami Chetanananda, with a lifetime of gratitude, ever and always.

Finally, my eternal thankfulness to S. N. Bear, Ambrosius Merlin, and Philip Hylos for their creative encouragement, heartfelt support, and unwavering guidance. It is their song I sing.

Gwendolyn D. Galsworth
Portland, Oregon
September 2014

PART I
The Cost Of Complexity

Introduction

Smart Simple Design/Reloaded is a book about how and why complication happens in businesses—and how to de-complicate, how to simplify.

The road from the complex to the simple is not a straight line. It is instead a spiral of logic, application, and thinking that defines and then dismantles certain aspects of causality. The broad paradigm for this progression is called improvement, even continuous improvement.

Currently, there are two powerful improvement methodologies helping companies. The first is *lean*, the second is *visuality*.

Nowadays, lean as a term has come to cover, like a huge mental umbrella, everything associated with improvement—from pull systems, concurrent engineering, kanban, leadership, employee engagement, and all things visual. By contrast, the actual definition of lean focused squarely on identifying and eliminating waste from the critical path—that stream of value (materials, information, and people) that results in products and services the company produces and the market will buy. Lean then is first and foremost a technical intervention that supports a specific, pre-determined outcome—install demand-based pull—no matter the attempts over the past two decades of well-meaning practitioners to stuff it with all things good and beautiful.

The second powerful improvement methodology, workplace visuality, serves its own specific purpose: to identify and eliminate information deficits (missing information) at work—information that is either incomplete, incorrect, late or not available at all. In a visual workplace, such deficits are removed when we implement visual devices because they imbed that missing information into the physical landscape. Suddenly, information is available as needed, no matter the moment, no matter the person. The result is flow—and a reliable and repeatable adherence to workplace standards, specifications, and requirements that improves safety, quality, cost, and on-time delivery.

Visuality and lean work powerfully hand-in-hand, much as the two wings of a bird work together—separately but in perfect balance and synchronicity. I have spent over 30 years working in the field of lean and developing the technologies of the visual workplace.

But when considered through the lens of the central theme of this book—variety effectiveness—visual and lean both fall short. Both can help us cope with complexity but neither can help us reduce its causes. In the parlance of this book, neither visual nor lean can achieve the *least cost means*—a splendid conceptualization of simplicity that addresses the entire cost spectrum. A line is laid from design concept to product delivery to bank deposit—the line of least cost means—and anything that does not hug that line is waste. Anything that does not hug that line is neither smart nor simple.

Despite the fact that the vast majority of companies reach for visual and lean to the deal with complexity, when they do they deal with its symptoms only, not complexity's causes. Visual and lean are premier operational improvement strategies but they do not—and cannot—dig deeply enough into cause to uncover and then eliminate the triggers of negative variety. In this respect, visual and lean represent limitations in concept as well as in application. They are adept at moving the furniture of complication around but not at changing the underlying architecture of causality.

The cure for complexity is found at a source deeper than an entangled value stream or a work environment overrun with information deficits. In the world of smart simple design, complexity is triggered on two levels. The first is the part. It is *the single part* that magnetizes problems, struggles, and costs. This plain cause-and-effect equation is both formulaic and reliable in terms of explanation and logic.

The second trigger is hidden in plain sight in company policy: both informal policies and practices that reflect the cultural preferences of the organization—and formal policies found in the rules and regulations that govern corporate practices. Unfortunately, these two triggers—the part-as-magnet and company policies—are rarely discussed in the literature or practice of either lean or visual.

Unaware of this, the company that reaches for lean and/or visual to handle its complexity woes will never attack the source of this waste, which is deeply rooted and nearly invisible. The organization may steadfastly implement visual and lean and achieve remarkable benefits in terms of time-, cost-, and distance-reduction. But it is only dancing around the issue of complexity, coping with—not tackling—a problem that is deeper and more insidious. No combination of visual and lean can ever solve complexity.

SSD/Reloaded endeavors to correct this misunderstanding and offer a methodology that can lead to a truly long-lasting solution. It does this first by describing the problem in enough scary detail that almost every reader can identify with the disease. No longer can we pretend that complexity does not refer to our enterprise.

Then it prescribes the antidote: VEP—a careful and complete methodology that is built on a robust conceptual base and a set of reliable tools and practices. The result is: the complexity that has gone un-noticed and un-tracked for decades is discovered and uprooted, replaced with principles, policies, and practices that bring the company back to life. Survival is no longer the goal. Profit is.

A Word About Additive Manufacturing

It is clear that the promise of additive manufacturing (3D printing) has grown in prominence and practicality as a revolutionary avenue for reducing parts and parts inventory. To begin with, this technology makes it possible to reduce the need to inventory or re-make parts for discontinued products. Even more exciting is the possibility to design and create parts with internal structures that could not be created by any other means—or even to combine two parts into one that could not have been done without this technology. In many cases, 3D printing makes it possible to combine parts, thereby reducing the number of parts used in existing products and more easily making them common between two or more products. Used for both prototyping and manufacturing, additive manufacturing can utilize a wide range of base materials, including plastics, metals, and, in some cases, human tissue. I urge you to research this remarkable development and assess its relevance for your enterprise. Although the magnitude of its potential impact on the issue of complexity and effective variety—as well as on associated support transactions and such ancillary costs as capital equipment and warehousing—has yet to be calculated, it is bound to be considerable.

CHAPTER 1

The Consuming Age: Relentless Pressures of a Voracious Marketplace

The management challenge today is to reduce costs—and increase the perceived value of the product. — ARTHUR L. KELLY, Deere, BMW, and NALCO Chemical

Started in a garage in 1963, Nike Shoes sold only one kind of running shoe during its first four years of business. Four years later, a second shoe design was added. Come the 1990s and Nike was introducing an entirely new product line—over 300 styles—every six months. Twenty years later, Nike produces more than 850 shoe styles *per quarter*, across eleven product categories. And Nike is not exclusively about shoes anymore—but also a proliferating array of golf clubs, athletic clothing, ice skates, baseball mitts and, well, what have you. Even horseshoes are proliferating, with over 600 different types—each with its own shape, width, and weight—and 50 kinds of nails.

In the fourteen years between 1913 and 1927, an automobile bought from the Ford Motor Company meant only one thing: the Model-T. Today buyers can select a Ford from among literally dozens of models and hundreds of options.

The opening of the former Soviet bloc nations, China, and other modernizing Asian countries has nearly quadrupled the world's available consumer market, with a potential increase in spending unparalleled in the annals of commerce. Apple initially sold 1.3 million smartphones worldwide in 2007—and by 2013 this had grown to 150 million. The marketplace is exploding in size, speed, and volume.

High-tech products are hitting the market like hot cakes off the griddle, with life cycles collapsing faster than anyone can track. A decade ago, you could count on a twenty-six-month product development cycle from concept to production. Today eight months is more the norm—but products can come into *and* go out of the market in three months. Like fresh fruit, they have an increasingly short shelf life.

Grocery stores and supermarkets in the United States can choose to stock their shelves from over 228,000 *new* retail products a year—adding to an average base per store of nearly 40,000 items, brand names and generic. The marketplace is exploding.

The evolution of disposable diapers is a good case in point. Throwaway diapers, which made their debut in the early 1960s with a single universal one-size-fits-all product, now come in dozens of differentiated designs related to—well everything: fastening tapes, absorbency, softness, layer thinness, waist and leg bands, decorative designs, and special padding "where boys and girls need it most." These features are further segmented by gender, age, size, and packaging. Each of the three leading makers, Pampers, Luvs, and Huggies, carries well over 50 product codes, with new competitors constantly making bids for Mom's dollars. Even Baby Wipes, which entered the family scene a short decade ago, come in dozens of brands and a myriad of styles: different scents, no scent, extra thick, all natural, biodegradable, flushable, and specialty wipes with added rash cream—all of it in a wide range of slightly different packaging.

Need more evidence of the exploded marketplace? In the early 2000s, Kellogg offered Eggo-brand waffles in 16 flavors. As of the writing of this book, that number has nearly doubled: 31 choices, this time including size. Similarly, Kleenex users were once satisfied with nine varieties of facial tissue; fifteen years later that has jumped to 49.

Product variations and collapsing product life cycles are the realities of the market. Companies previously tied to a tight linear new product approach are now learning to develop complex roadmaps of overlapping product generations—for many generations ahead. More than at any previous time in the history of business, the market is driven by the consumer's demand for choice. Customers are in the driver's seat—and they know it. They are voting with their pocketbooks in unprecedented numbers. And companies are scrambling to capture them with expanded products.

Competing through new products has created a whole new set of rules. Winning today only means that you get a chance to play in the next round tomorrow. This is, however, not just a time of widening consumer choice—challenge enough for companies. The fact that the customer is king is the good news. The sobering news is that this is also a time of low pricing.

In the 1800s, product availability created the marketplace: If you could get it there, you could sell it. By the turn of the century, price or affordability defined the public's buying patterns, only to be replaced 20 years later by choice or product differentiation. In the 1970s, quality or performance surfaced as the principal consumer yardstick. Then, in the late 1980s, quality linked up with service for the promise of total customer satisfaction. Now what consumers want—and often get—is total availability, choice, quality, performance, and service. But there is more: They want these at ever-shrinking prices. The age of disinflation has arrived.

The Age Of Disinflation Has Arrived

Disinflation refers to a marketplace phenomenon driven by twin forces: rising product value and falling product prices. Formerly, a product's selling price was a function of cost plus an acceptable margin of profit. In the boom of the 1950s and 1960s, a business always had the option of absorbing its cost overruns and improving the bottom line by simply hiking up the price tag. Inflation—the name of that game—covered up many sins. Those times have passed. After decades of marking up prices to cover inflationary rises in cost, manufacturers find themselves forced to reverse their customary pricing practices. In the face of disinflationary pressures, they must lower the price even as their costs and consumer demand soar.

Disinflation is an anomaly. A rush for products normally generates pressures for higher, not lower, prices and produces higher, not lower, inflation. This time, however, the opposite is happening. Traditional price setting (price = cost + profit) is dead. Instead companies—using a strategy known as price-targeting (also known as "design-to-price")—begin by setting a target price for a new model based on the customer's perspective, then back down into cost. The pressures to cut costs become extraordinary.

The message is clear: Your opportunities in today's turbulent marketplace are truly unprecedented—*if* you can get and keep your costs down. Unfortunately, many organizations are blazing trails into this new competitive wilderness armed with weapons from the dark ages. Not only are they continually exposed to attacks from low-cost rivals, but their own business practices are making their survival, let alone their success, very iffy.

The Challenge

In this new world of disinflation, cost-cutting is, of course, essential. The price a company charges is the culmination of every decision it has made along the line. Without the cloak of inflation, all those decisions are directly exposed to the ruthless pressures of the marketplace.

Disinflationary forces are setting a new cost cutting agenda for even the most successful companies, a challenge made even more demanding due to high levels of product segmentation and international competition in the marketplace. Organizations that are leaders in product variety and product innovation stand to be the greatest losers. The very variety that won them markets during inflationary decades could now sink them. Many of these companies do not realize that they could become victims of their own growth.

The fact is: Even when design-to-price products create a consumer demand that makes a company's market presence unassailable outside the organization, the same products can create complication and confusion on the inside. Supporting an endless stream of varying products triggers a chain of events within the enterprise that burdens its infrastructure to the breaking point.

To a large extent, this torrent of new products is the stepchild of widening capabilities in computer technology, at best a double-edged sword. One edge gives a company the ability to churn out new products in unprecedented number and variety, offering customers a selection and allowing the company entry into markets previously beyond reach. But the same computer revolution has released a flood of complexity. Leaps in product variety steadily and inexorably swell parts inventories and production processes. The myriad supplementary activities that support these are similarly geometrically multiplied: material handling, dies, tooling, fixtures, changeovers, maintenance—as well as IT/data systems, drawings upkeep, and all underlying transactions (aka, *control points*). This buildup happens gradually—product by product and part by part—as the overall system makes a series of barely discernible micro-accommodations. The complication and congestion accumulate and the company begins to grind to a halt. It simply can't handle the detail. This is the cost of complexity.

Even when sales soar, the parallel expansion of organizational costs, complexity, and congestion can choke the infrastructure. Profits take a nosedive and the company founders. Attempts to maintain profits by cutting headcount, travel budgets, discretionary funds, and other overhead expenses not only miss the point but are often implemented without a plan of how fewer people will maintain—let alone improve—current levels of performance in the face of the big squeeze: sizably reduced resources and over-stretched, harried, and tired employees. The smart organization, by contrast, understands the dangers of unchecked variety proliferation and takes concrete steps to either prevent it—or, if already in the throes, get it under control and then reduced.

The Iceberg Effect: Hidden Danger

Like an iceberg that has two-thirds of its mass underwater, most of the negative consequences of mushrooming variation are hidden from view (Figure 1.1). Many companies that appear to thrive are, in fact, on their way to a crash because they lack the know-why and know-how to prevent or control these disastrous effects. Whether they go belly-up quickly or sink slowly and painfully, it will be because they mis-identified the problem as deteriorating profits or excess inventory—and, therefore, the solutions they apply do not address the real problem. Whatever salutary inroads may be made through the methods and strategies of lean production and continuous improvement (and they are considerable), these will not help the company confront the problem *underneath* the production system. The problem is not the production system. It is not quality, service or delivery. And it certainly is not the customer. The problem lies within the product itself.

Figure 1.1. **Danger! The Real Problem Remains Hidden**

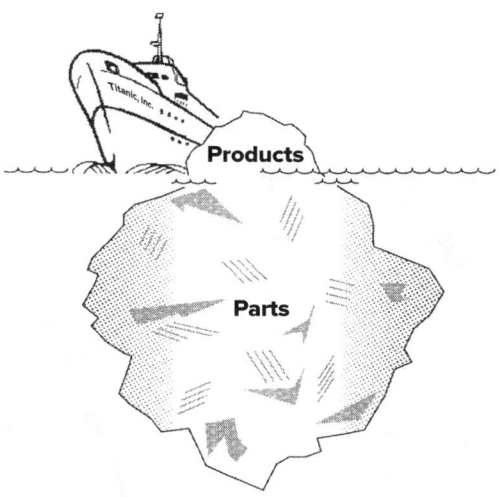

Flawed assumptions about product cost and the effects of product variation on profit, complicated product structures, runaway part numbers, a computer classification system that obstructs engineers from easy access to existing part specifications—all conspire to make the rush for new products tantamount to corporate suicide for many maturing companies. Unfortunately, many young, up-and-coming enterprises may also be on their way to a shipwreck.

The globalization of the marketplace has triggered unprecedented levels of opportunity—and, with them, complexity that can topple the brand new as quickly as the old. Complexity's impact is always a geometric—never a linear—expansion. It is therefore crucial for you to learn to define what is positive about that impact and what is destructive. You need to learn how to define and determine the cost of complexity.

Variety Effectiveness

The purpose of this book is to inform organizations of the dangers of uncontrolled proliferation, delineate its symptoms, spell out causes, and prescribe a cure. That cure is *VEP: The Variety Effectiveness Process*—a systematic process for preventing or reducing unwarranted variation while offering maximum customer selection at a least-cost sum.

VEP deals directly with the front-end interface between marketing and product design. It works to de-complicate the organizational infrastructure and cut total costs while developing exceptional selection for customers. Whether proliferation is in parts, products, services, SKUs, market offerings—and the names of those offerings—the synergistic, multiplying effect of *more and different* is the cause.

VEP includes a systematic assessment of the company's market offerings, the structure of its products, and the range of variation within each type of part in its parts inventory. In implementing VEP, a company can reduce its parts inventories by as much as 40% by modularizing and streamlining product structures and by eliminating or simplifying parts or making them multi-functional. As a direct result, the company also sees impressive reductions in the number of production and supporting processes, in associated tooling, fixturing, and dies, and in the paperwork and computer transactions that support all phases of the product development and production process—as well as in associated sales and marketing functions.

Through specific step-by-step procedures and a range of hands-on tools, VEP helps companies find and eradicate the causes of complexity—unwarranted variety in engineering, marketing, IT/data, purchasing, accounting and support, and operations. For both mature and start-up manufacturers (or any enterprise for that matter), the bottom-line impact of the Variety Effectiveness Process can be dramatic and long lasting.

VEP primes and prepares the organization to grapple with the real challenges and opportunities of the marketplace. In this period of exploding consumer demand and disinflationary pricing, VEP answers the central question of every organization: How can we continue to capture customers with products and services they genuinely delight in, at a *cost* low enough to set a *price* low enough to compete—*and* still make an acceptable profit year after year, as the company continues to grow? This book answers that question.

The VEP process is powerful. Even more powerful is the concept of *variety effectiveness*. That's what companies really have to understand. Simply introducing exciting new products is not enough. A commanding market share is not enough. To make your company consistently more profitable, the variety you offer in those products to those markets has to be effective.

Effective variety is the balance point between variety that adds value in the eye of the customer (positive variety) and variety that merely adds cost (negative variety). The first is customer-driven; the second is internally triggered. Effective variety is a continually shifting balance point, moving more and more toward the positive pole as the company lessens its load of negative variety and augments the positive kind.

The result is *smart simple design*. *Smart* because it capitalizes on the best of the company's existing designs and parts, while standardizing, modularizing, and integrating as much as possible—even as it maximizes customer selection. *Simple* because it confronts unneeded variation in products, parts, and processes and systematically attacks policies and practices across organizational functions—specifically, Accounting, Purchasing, Design Engineering, Operations, and IT/data systems—that promote negative variety and eat away at profits. And *design* because variety that is negative can only be prevented at the concept stage. *Smart simple design* is the answer to the question: How do we build profits in the face of widening customer demand for increasing product value at diminishing product prices?

About This Book: Chapter By Chapter

The goal of *Smart Simple Design/Reloaded*—and the VEP Methodology it presents—is to help you identify opportunities to simplify your company, dismantle the levels of complexity that have taken root there, and, once done, prevent their recurrence.

Smart Simple Design/Reloaded is structured with two different audiences in mind. The first audience is CEOs, CFOs, and VPs. The message to them is: Negative variety is killing your company—and change is needed. The second audience is managers, middle managers, designers and engineers, operations managers, marketers, and finance people—people who deal with the everyday effects of decisions made at the top and who are expected to implement those decisions and pro-actively mitigate against unintended consequences—in other words, improve those decisions.

The book is segmented into three parts. The first—*The Cost of Complexity*—describes the origins of the problem and why it is important to do something about it. Chapter One talks about the turbulence of the marketplace and the urgent need to cut costs. Chapter Two looks at the symptoms of unwarranted variety—and the fact that most current improvement strategies can have only a limited impact on the cost of complexity. Chapter Three discusses how traditional GAAP cost accounting can mask negative variety and then presents VEP's approach to this, including its *Tri-Cost Model* and *Parts Index*. In Chapter Four, we examine formal and informal policies that provoke negative variety. In the fifth and final chapter of Part One, we suggest ways designers and engineers can use the principles of variety effectiveness and build them into products that sell and delight.

The second part of the book, *VEP: The Methodology*, describes VEP's systematic process for identifying and reducing proliferation and presents specific tools and procedures for preventing or minimizing this and preventing the recurrence of negative variety. The first chapter in this section (Chapter Six) describes how to set up the organization for a successful VEP implementation—structuring VEP Teams and selecting a valid starting point for VEP analysis. Chapter Seven details a procedure for achieving the crucial prerequisite for success: a VEP-capable classification system. Chapter Eight presents the set of six tools that focus VEP analysis on tangible, measurable improvements. These tools find their context in Chapter Nine where we share the approach called *3-View Analysis*—

VEP's step-by-step process for identifying and reducing negative variety in marketing, product structures, and parts types. Next, Chapter Ten applies the 3-View Analysis process to the reduction of processes and control points. Chapter Eleven, the final chapter of Part Two, rounds out the discussion of the VEP method by explaining the steps for coordinating, prioritizing, scheduling, implementing, and sustaining improvement proposals.

Part Three, *The Bottom Line* (Chapter Twelve), returns us to the big picture: the cost of complexity and its imbedded hold on the organization—and how to release that hold through variety effectiveness mindset. The focus is squarely on the overall goal: increased profitability and the dangers of ignoring the challenge of effective variety. If decision makers in your organization are looking for more evidence of VEP's impact, this short chapter makes a compelling argument. At the close of the book, you will find a glossary of terms and an index.

Variety effectiveness is a vision of the change to come—and the VEP Methodology is designed to achieve that vision. In *Smart Simple Design/ Reloaded*, vision and method are joined to provide you with the insight and understanding you need to choose the change—and the knowledge and skill to realize variety effectiveness in your organization. Our wish is to leave you not just with a sense of urgency about the problems hidden in product—and service—expansion but also with the conviction and ability to tackle them.

Smart Simple Design/Reloaded

CHAPTER 1

CHAPTER 2

On The Horns Of The Dilemma

There is at least one point in the history of any company when you have to change dramatically to rise up to the next performance level. Miss the moment and you start to decline. **ANDREW GROVE, Intel**

Several years ago, I was asked to meet with the president of a medium-sized electronics company, an actual client I worked with for a number years that I will call Parts Unlimited Inc. (PUI). With names and certain other details modified, the PUI case study is threaded through the remainder of this book, providing a fine way to examine the impact of unwarranted variety, holistically, within a congruent organizational setting.

The purpose of that initial meeting was to analyze a troubling condition the company was experiencing: increasing revenues and plummeting profits. The VPs of Engineering, Marketing, and Operations, along with a number of company managers, were scheduled to attend as well.

PUI: Exponential Growth/Less Profit

Privately established in 1932, PUI entered the electrical switch and control business with a single breakthrough product at a time when the market for such products was just beginning to expand. An instant hit, the new product became a nationally-recognized brand within two years. In the same two years, the workforce tripled, as did the square footage of the facility. Annual sales climbed from $1.2 million in 1935 to $9 million in 1940. Since then PUI has remained an industry leader. Although the company faces staunch competition from look-alike products, it still holds sway over an impressive 67% of market share.

It was in response to the military's need for new electronics-based products at the start of World War II that the company's product offerings exploded. Using its core products as the base (at the time there were five), PUI engineers added more than 75 new models to the company catalog by the end of the war. In 1948, the company launched an aggressive product expansion strategy which, 15 years later, had created a total of 18 product lines and more than 200 models as well as hundreds of customer-specific options. Possible combinations numbered in the tens of thousands.

Ten years later, specific model numbers hovered near 400, and, by the late 1980s, partially as a result of the new market-in design efforts, that number had doubled again. Every year since, PUI customers have seen yet more new products. A few of these are highly innovative, pushing the industry to the next level. But many are derivatives, improvements, flankers, and extensions—reconfigurations of existing products, only slightly modified to respond to the company's ever-multiplying market niches.

As a result, PUI has earned a name for itself over the years as a customer-driven organization. Its quality is high, response time competitive, and prices attractive. The company is considered one of the hottest lean plants in the region and a model of operational excellence, cell design, kanban, quality function deployment, and visual systems. It has won many awards for its improvement efforts and continues to search out breakthrough technologies in its never-ending quest for new and larger markets, lower costs, and greater profits.

Today PUI, still privately owned, offers more than 80 lines of products and well over a thousand models, plus countless options. Fifteen lines are the company's key money-makers, accounting for nearly 65% of the revenues. Another 20 lines contribute about 25% in yearly receipts. Demand for the remaining 45 product lines is sporadic at best—but, as yet, not so infrequent that any have been withdrawn. In addition, never wanting to disappoint an established client, the company has maintained a replacement policy for any part or product it offers—or has ever offered.

Visual Evidence of a Problem

Crossing the shop floor on my way to the 9:00 a.m. meeting in the Engineering Department, I pass through various operational areas. Stacks of point-of-use storage bins overflow with thousands of small parts; subassemblies line the racks on either side of the aisles. Shelves along the way are covered with tooling and fixtures, all neatly tagged, color-coded, and arranged. Small quantities of work-in-process (WIP) are stacked in orderly arrays in each work area.

I detour through a maze of desks where schedulers sit sifting through various Bills of Material (BOMs), preparing to get new orders rolling and stage the old ones. A dozen or more expediters are gearing up to handle the new schedule, the best of whom have thirty years with the company under their belt; they know how to work the system. As I turn down the hallway and head toward the conference room, I bump into Harvey Chasewaite, first-shift foreman and veteran of forty years at PUI; he's muttering to himself something about missing kits.

Finally I reach the Engineering Department and the meeting room. Piles of drawings and other documentation cover every available surface. Work in the department has a purposeful air. On entering the meeting room, I see Tom Vargas, PUI's president, at the head of the table surrounded by his direct reports. People are chatting in twos and threes as they wait for the session to begin. As I pass, I overhear Gary Scosberg, VP of Marketing, expressing doubts to the chief purchasing agent, Gerry Beryl, about the Engineering Department's ability to meet new product introduction deadlines in time to coincide with the new marketing catalog, due in five months. Scosberg's voice is good-humored but his face looks drawn and tired. "We're not talking rocket science here," he says. "We just want a few new features! Why can't people just design some simple extensions without leaving their personal creative paw prints all over them?"

Camilla Wardwell, VP of Operations, appears equally exasperated as she converses with Vargas, who seems intent on every word. I draw closer and hear her saying: "I know we can't just slash and burn our product line into manageability, Tom, but this customer-driven thing is getting out of hand. My expediters don't even know which of our products make real profit—and neither do I. If we introduce one more new product line this year, my staff will lynch me."

Dan Littel, Engineering VP and 30-year veteran of the company, is sharing a similar anxiety with the finance director, Maureen Gleason, recently promoted from the ranks of Purchasing: "We can't handle any more product, Maureen. Did we waste all that time and money on lean? I don't know exactly what caused it, but I do know that something is out of control and we are heading for a slide."

At the stroke of nine, Tom Vargas rises to his feet to begin the meeting. His opening sentence shocks the room into attention: "My company has become my own worst nightmare." After a long pause, he continues: "Our markets are turbulent, profits have steadily declined over the last decade, and we're looking at an eroding market share for the fifth year in a row. But our inventory continues to soar, up from $4 million three years ago to $5.6 million this year. In exactly the same period, our active parts count mushroomed from 9,454 to 13,156. That, ladies and gentlemen, is an average increase of 6.8% per year!

"We've been over these figures before. I have asked for and heard your plausible explanations: To get more market share, we need to offer whatever it wants—and new products require new parts. Customers want things fast so we need the inventory to cover. And, oh yes, our suppliers' lead times are horrendous.

"Well, I just met with Andie Randal and she gave me another piece of news: IT anticipates that our inventory will hit $6 million by the end of the year if our current rate continues. And Dan's group is predicting parts count will go up 20% if his engineers complete all the ECNs [engineering change notices] and development projects on their desks as of today.

"Let's not kid ourselves. We all know that this is just the tip of the iceberg. There is something else out there. Walk around the shop floor. Spend some time in Drafting. Hang out in Purchasing for just a while. Talk to our sales guys and gals. Go to the machine shop. Listen. It's out there. This company is choking on its own chronic busyness. You can see it and you can feel it. And I want a name for it. And I want to know what we're going to do about it. But first I want to know why. What went wrong?"

What Went Wrong?

From its earliest days, Parts Unlimited Inc. dedicated itself to providing its customers with anything and everything they wanted. As a result, the company has been well-rewarded with a loyal and ever-widening customer base. Now, sixty years later, the company continues to dominate the market. Business remains brisk and revenues increase every year. But the company is in trouble. Even though sales volume is up, profits are down. They haven't just leveled off—they are shrinking. For the first time in the history of PUI, break-even has become a critical issue. Despite the organization's considerable success in waste elimination, cycle-time reduction, and continuous improvement, costs are eating the company alive, and managers can see nothing but red on the horizon.

There are many companies like PUI. After introducing a distinctive and popular new product, they grow by leaps and bounds during their early years. When competition rears its head, such companies often confront the problem by diversifying their product offerings. If the strategy works (and it often does, short-term), product expansion continues as these organizations pursue higher sales volume, greater revenues, new market share—and, of course, more profit.

But, like PUI, these same organizations sooner or later find themselves in the same puzzling "more revenues, less profit" conundrum, coupled with the operational gridlock Tom Vargas referred to as "chronic busyness." The search for the culprit and solutions begins anew. It is often at this point that cost-conscious companies launch inventory reduction initiatives. If company managers are well informed (and the vast majority of them are), they will select an array of the powerful and highly effective techniques clustered under the rubric of lean, continuous improvement or waste elimination.

Before examining the real source of the problem challenging PUI and companies like it, let's review the above-mentioned techniques to better understand the impressive and positive impact they can assert on a troubled enterprise—and their limitations.

The Improvement Revolution

In the late 1970s, savvy companies began a transformation that is still ongoing. This transformation focused squarely on replacing antiquated operational approaches with dynamic new methods that improved quality and productivity performance, accelerated the flow of products and services to the customer, and helped companies become superior competitors. The key to the revolution was (and remains) ridding all systems of waste. *Waste* was the culprit, and waste reduction was the strategy for eradicating it.

In this approach, waste in a company is broadly described as any *activity that adds cost and not value to products or services*. A synonym for waste is non-value-adding activity (NVA); and its opposite is known as value-adding activity (VA). Together, value-adding and non-value-adding activity make up the total of a company's endeavors:

| VA | + | NVA | = | TOTAL COMPANY ACTIVITY |

Research shows that an average of 95% or more of this activity is non-value-adding, leaving only about 5% that adds value. Said another way, for every 100 hours of "work," 95 hours (on average) are spent in doing things that do *not* add value to the product or service, with only the remaining five hours spent on activity that *does* add value.

Non-value-adding activities fall into seven broad categories of waste:

1. making defects
2. delays
3. over-processing
4. motion
5. overproducing
6. making inventory or WIP, and
7. material handling.

Examples of non-value-adding (NVA) shop floor activities include: double handling products, searching for tools, long equipment changeover times, waiting for parts, waiting for instructions, and unscheduled equipment downtime. Examples of value-adding (VA) activities on the shop floor include forming, stamping, machining, drilling, heat treating, assembling, and painting. Add to these *missed opportunities* (opportunity loss) from using limited resources to support waste, and you have the *Seven Deadly Wastes + One* (Figure 2.1).

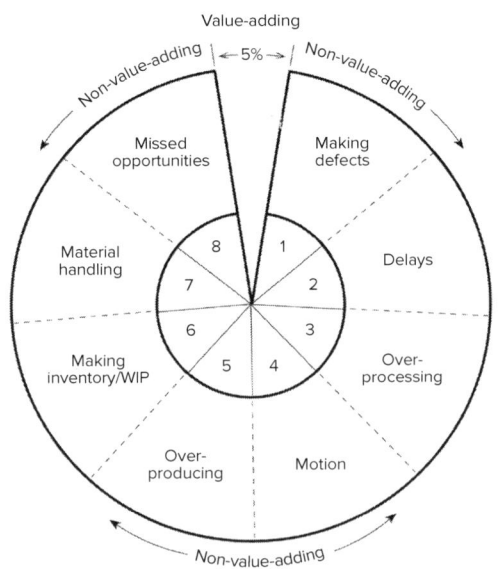

Figure 2.1. **The Seven Deadly Wastes + One (Non-Value-Adding Activity)**

These wastes are often referred to as rocks, blockages or debris in the river (the river representing the flow of product through the production system). Only the tops of them are visible above the water line; the rest is masked by the enormous inventories a company keeps to buffer against unanticipated sales spikes and such production demands as defects, equipment breakdowns, long equipment changeover, etc. (Figure 2.2). Getting the product to the end-user on time is not an unobstructed flow down the river but rather a painfully slow and sluggish meander around the rocks that choke the flow.

Figure 2.2. **Rocks in the River: The Flow of Production**

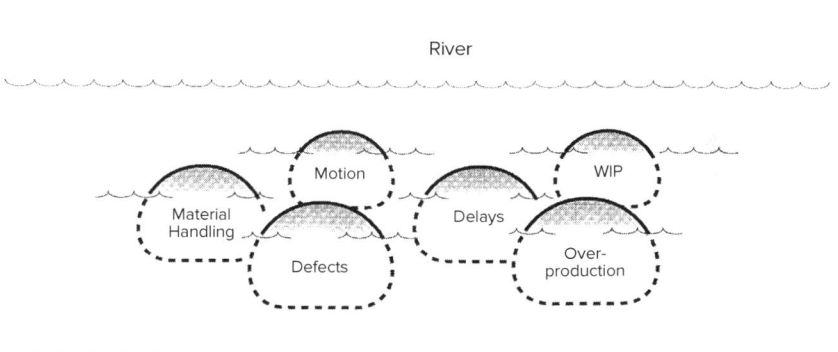

The objective of the waste reduction strategy is to systematically identify and remove those rocks or wastes (the NVA component of the formula), including the inventories themselves. A key element in this attack requires managers to resist "improving" the value-adding (5%) component (Figure 2.3). Their exclusive improvement focus needs to be on the 95% that represents waste.

This strategy was raised to both a science and an art by Japan's Toyota Motor Company in the 1970s and has since become the focal point of the improvement efforts of many companies all over the world. Today, definitions of the seven deadly wastes are practically burned into the brain cells of self-respecting executives, managers, and shop floor employees everywhere—along with the methods and techniques used to reduce these wastes and shift the VA/NVA ratio into greater balance.

Quick changeovers, kanban, statistical process control, fail-safe/poka-yoke devices, cell design, supplier-customer partnerships, visual workplace technologies, quality function deployment, and other powerful

Figure 2.3. Value-Adding Activity vs. Non-value-Adding Activity

---

Value-Adding Activity vs. Non-Value-Adding Activity

When asked to describe what it "does" all day, a company will often reply: *"We're busy all the time."* The company equates busy-ness with work and sees that busyness as 100% value-adding (VA).

| 100% "BUSYNESS" | = | WORK | = | VA (VALUE-ADDING) |

Based on extensive research across many industries, the truth of it is more likely this: 95% of the activity is non-value-adding (NVA)—and only 5% of the activity is actually value-adding (VA). That is, for example: Out of 100 hours, only 5 hours represents actual work, with the remaining 95 hours simply waste. This is known as the "95/5 non-value-adding/value-adding ratio."

| 95% NVA (NON-VALUE-ADDING) + WASTE | 5% VA |

When a company is not aware of this 95/5 ratio and wants to improve its profit margin, it often attempts to improve that portion of activity that adds value (5% VA)—instead of attacking activity that is non-value-adding (the waste/95%/NVA). For example, this year ACME Inc. targets a 100% profit improvement over last year and conceptualizes the growth something like this: "Last year we made and sold 50 widgits. This year we'll make and sell twice as many widgits—100 widgits—and thereby double our profit. For sure!"

Based on this mistaken logic, ACME Inc. proceeds to improve its VA portion—buying more machinery so it can make more product. But because the relationship between VA and NVA is a ratio, making and selling more widgets will automatically cause the VA portion to shrink and the NVA portion simultaneously to increase. As a result (in this case), the 5% VA shrunk to 2.5% VA while, in parallel, the 95% NVA expanded to 97.5% NVA.

| 97.5% NVA | 2.5% VA |

On the other hand, if ACME Inc. had understood the logic of the 95/5 ratio, the company would have first focused on reducing the volume of its non-value-adding activity so that NVA would diminish. Only when the ratio between VA and NVA reaches parity (50%:50%) should a company focus on improving the VA portion.

| 50% NVA | 50% VA |

---

continuous improvement techniques have entered the mainstream of the way companies now do business. And they are working. In many companies, processes have been streamlined or completely re-engineered, work-in-process (WIP) has been drastically reduced, lead times have shrunk, inventory turns have improved, single-minute changeover times have become commonplace, and transportation distances have collapsed. The non-value-adding side of the ratio has receded.

On the value-adding side, the demand for new products and market-in design has quickened. Many manufacturing companies face this reality: Successful new products are the lifeblood of the enterprise. Today, zero defects quality, nearly instantaneous delivery, and broad product selection are virtual givens in the marketplace.

Make no mistake. These approaches are genuinely powerful, even transformative; they have delivered on their improvement promises. The results have been dazzling by any measure. But all of them, without exception, start on a secondary level of cause. None address the underlying condition that spawned them. What celebration and relief when Beowulf slaughtered Grendel. But more was to come: Grendel's mother surfaced from the primordial ooze. The fundamental circumstance that triggers the need for lean, visuality, TPM, six sigma, and other improvement initiatives common in the field of operational excellence is *negative variety*. The current alphabet soup of improvement tools is simply an array of coping mechanisms for that.

The fact that negative variety is rarely seen or named as culprit is of little importance. Its invisibility does not diminish its destructiveness. Please bear this in mind as you contemplate your next improvement investment. Will you resource yet another improvement tool aimed at making smooth the symptoms of negative variety? Or are you ready to attack the source? That one new part? That one new name for a product offering? That one new model? That one new service? There is a tremendous requirement of all businesses today: new products. Every business—mature or shiny new—must eventually face the challenge of product glut.

The Push for New Products

Ours is an age of choice. We live in a time of abundant products. And the customer is hungry. It wasn't always that way—but it is now. The past two hundred years have witnessed a revolution in the industrialized world of what, why, and how people buy—the birth and ascendancy of consumerism.

In the late 1800s, constrained by rudimentary production techniques and poor roads, most Americans bought goods that were locally made and locally sold. This was the epoch of commodity wares, general products that met the practical needs of the day—boots, clothes, carriages, and "isinglass curtains that'll roll right down in case there's a change in the weather." Availability was the defining competitive characteristic.

As national transportation networks evolved and production technology improved, companies extended their reach. The buying public gained access to a wider choice of wares and higher-quality goods, uniformly made. This was the era of Henry Ford's great contribution: the assembly line and the standardization that made it possible, an extraordinary leap in concept and execution. Think about it. Ford tested his first automobile in 1896, four years before the national census reached 76 million Americans and 21 million horses. By 1903, 11,235 cars had been sold in the United States, a number that jumped to 43,000 by 1907. The year 1908 was to become the watershed between *commodity* and *standardized* products—between the purchase motivator of *availability* and that of *affordable sameness*.

That was the year Henry Ford introduced his famous "any color as long as it is black" Model T, the ultimate universal product. In an era when one product was perfect for everyone and price defined the competition, Ford's Model T became a national brand, the ultimate perfect-for-everyone product. That 1908 standard remained unchanging and unchanged for nearly two decades until the Model T was retired in 1927, with over 15 million sold. The age of universal products reached its peak during World War II, with its urgent need for highly standard, mass-produced goods.

As a result of that war, the public had grown more discerning, more sophisticated, and wealthier. People wanted special things, products that reflected their personal life-style and values. Product diversification was born. What started back in the 1920s—when General Motors debunked Ford by introducing the concept of different cars for different people ("a car for every purse and every purpose")—turned into a headlong rush of annual model changes, market segmentation, and products that were differentiated on the basis of ever more finely delineated sets of demographics. Americans, above all other people, had come to cherish their right to choose. The day of the consummate consumer was just around the corner.

That corner was turned in the 1980s with the advent of mass customization, a customer-driven strategy that combines the best production systems with a never-ending series of product choices. B. Joseph Pine II, one of mass customization's leading advocates, makes no bones about it: The way to create products that sell is through optimizing relationships with your customers. He goes on to say:

> *Mass customization is first of all a mindset that places the needs of each individual consumer paramount. It holds that no customer should have to sacrifice what he or she needs and wants because of a company's internal inability to provide it.*

Once you grasp that mindset, you are ready to consider how to go about differentiating and customizing your products—and doing it at low cost. You generate a strategy and then, still holding fast to that fundamental customer mindset, you develop the technology you need to be able to deliver it.

Once again the auto industry is a perfect reflection of new trends. In their admirable and exhaustive treatment of the global auto industry, authors Clark and Fujimoto observe: "Where [40] years ago the American car buyer had to look long and hard to find a model with anything but a traditional V-8 engine with rear-wheel drive, today choice in engine drive train spans 4, 6, 8, and 12 cylinders, multi-valves, front-wheel drive, and 4-wheel drive." Since then, more variations in electronics, interiors, and brake, suspension, and engine control systems widen customer choice even further. Similar to the disposable diaper market discussed in the previous chapter, the range again expands exponentially when automobiles are segmented psychographically. You get: passenger cars, hatchbacks, luxury sedans, economy cars, sports cars, sports coupes, station wagons, vans, mini-vans, SUVs, CUVs, recreational vehicles, and so on and so forth. More variety than you can—or want to—shake a stick shift at!

Whether or not customers really want all those choices is another story and one we will address later in this book. Right now, let's look at other forces driving product variety.

Expanding Choices—Collapsing Cycles

Businesses cannot survive, much less prosper, without continually bringing out new products. Few companies can rely on markets secured ten or twenty years ago—or even one year ago. Whether to respond to a new consumer demand or create one, new products drive markets, and companies dedicate enormous time and effort each year to introducing them. Some new products are simply extensions, flankers, improvements or "me too" entries. Others are new to the company or even new to the world. Some sell and some fail.

Whatever the case, it is no longer enough for a product to be of the best quality, delivered the fastest, and at the lowest price. In a marketplace where winning and keeping customers is the name of the game, companies must offer all this in addition to the widest possible choice of products. Consumers expect and demand greater convenience, higher performance, and an ever-widening variety at equal or greater value—more features that are personalized to them. These requirements drive the market and the pace is accelerating.

Contributing to the speed of product proliferation are two other factors—collapsing time-to-market and product life cycles. There was a time that it could take anywhere from three to five years, in most industries, to get a new product introduced. Other industries, such as automobiles and other heavy equipment, could require more. Now, more often than not, you or your competitor, aided by computer-aided design and manufacturing within a concurrent engineering approach, can get the job done in months. Once the new product is launched, productive life expectancy, which was formerly decades, is now also counted in months—and is shrinking.

The benefit for the consumer is practically unalloyed—an endless stream of exciting new things to buy. But for the manufacturer, accelerating time-to-market and collapsing product life cycles act as a double-edged sword. One edge represents a significant competitive advantage (if the company is positioned to take that advantage). On the other edge are products that come and go so quickly that a company has less time to recoup its development investment, justify product costs or reach important economies of scale. And when products fade and are retired, the cycle continues and these products get swiftly replaced by "new and improved" ones. The net effect of these factors is an ever-widening spiral of variety, a virtual explosion of products.

Exploding Variety: One New Part

There is no question: To prosper, companies must develop successful new products, products that offer distinct and meaningful points of difference in the eyes of the customer. But "successful" does not only mean valued, wanted, and bought by the consumer. It must also mean greater profits for the manufacturer. To be a true success for the company, new products must represent a *least-cost sum* (achieving maximum customer selection with the least amount of resources or cost).

Too Much of Too Much

What companies often fail to see in their rush for more new products is the increased stress each new product adds to the company. The first sort of stress derives from the strain on resources required to develop new products. New consumer products can cost one million dollars or ten million to get from concept to final prototype, with associated production technology and marketing eating up millions more. The stress on the organization of this repeating level of investment is significant and observable.

There are, however, other less evident categories of stress beyond those generated by the cost of development which a company rarely even notices: the stress of adding one new part.

Estimates on the cost to develop, release, and carry a new part in inventory are hard to come by, as further evidence of the lack of focus—or even awareness—in this operational aspect as a cost factor. One survey of 18 companies and 30 divisions found that the introduction of a single new part cost the organization between $123 to nearly $6,000—an average of approximately $1,500 per new part. Engineering costs associated with that part averaged around 20% of the total cost. When you multiply the impact of that single part occurrence by the host of parts that enter a system annually, it should be enough to make a company sit up and take notice. But first they have to look.

This point becomes even more dramatic when we see that this survey, which spanned nearly 100 manufacturing plants, was conducted between 1967 and 1968. Today, studies estimate that the average cost of introducing a single part is not less than $4,000 per part, conservatively speaking— and usually requires weeks, even months, to bring it into actual production. The same studies calculate that designing a single part can represent as much as 20% of the product design cost or more.

As importantly, the life of that part and its associated part number can range from two to 20 years. What then, we must ask, is the cost of that part over its life cycle as a component of the company's inventory? Figures that scratch the surface of this cost are calculated at $60,000 to $100,000 per part. But, again, the marked scarcity of research on this factor means that few organizations prepare to harvest the cost savings from careful control of new parts introduction. This robust opportunity is simply overlooked.

Venkat Mohan, president and COO of CADIS Inc., once a leading parts-management software/system vendor, rightly commented: "The bulk of the part-life-cycle cost is in soft hours and is not easily measured using conventional systems that focus on hard dollar costs. And, the responsibility for managing these enterprise-wide costs is often undefined." CADIS, one of the lone voices in the 1990s that attempted to rally us to the need to examine part-life-cycle costs and implement ways to reduce them, has all but vanished through multiple acquisitions. We know of no firm that has replaced it in passion or precision.

When a company introduces a new product, you can be sure that at least one new part gets added to the parts universe. Does this surprise anyone? Certainly not. "What? Only one new part!" you may say. "When we come out with a new product, we add anywhere from 20 to 200 new parts to our base! What's the big deal of only one new part?"

The big deal is that every single added part puts a new burden, however small, on the organization. Let's look at what happens in the wake of introducing one new part, which we define as: *a part that is not currently in use or available inside company walls.*

In the wake of the addition of a single new part to a company's parts universe come legions of secondary activities. At the low end, a new part triggers the need for at least one new drawing. Then comes the need to contact a supplier (or possibly find a new one). Contacting a supplier triggers a purchase order, various phone, fax, and computer transactions, and at least one check. The part then needs to be handled in some manner, which may include some or all of the following: receiving, counting, inspecting, shelving (whether point-of-use storage or in the stores—that is, if space can be found), and when the part is summoned, it gets handled again. There may be some other "minor" requirements as well. If the company is lucky, for example, the new part will need only one new tool—a special wrench, perhaps—and only one new operation sheet or procedural write-up. But it might also require the purchase of new fixturing or even a new piece of equipment. In either case, in addition to the purchase cost, another cycle of paperwork and computer transactions gets triggered. And then there is the question of the added load on the production control schedule.

In and of itself, each of these activities could be said to be individually negligible. As a whole, however, they trigger stress in the organization that is observable and problematic, over time creating a tight web of actions that adds complexity, complication, and cost.

The Eight Runaway By-Products of One New Part

Just as specific waste categories were named (above) as burdens to traditional manufacturing operations, so too can we define and label the excesses or wastes associated with new product introduction. In effect, these wastes are the internal *by-products* of new products—and of adding one new part. When a certain level of critical mass is reached, these get the upper hand and "run away." For that reason, we call them: *The Runaway By-Products of New Product Expansion.* They fall into eight broad categories:

   1. Exploding Active Parts Count. When new products are developed, new parts get added; the question is: Is each of these new parts required and unavoidable? Even if a new part is required, it may bring an escalation in the number of service parts the company must stock.

   2. Pressurized Sourcing and Procurement Activities. As a part number is added, sourcing and purchasing teams respond; the continuous need for new purchased or made parts exert pressure on the parts procurement function as well as on other related functions.

   3. Unwarranted Processes, Dies, Tooling, Fixtures, Equipment, and Changeover Times. New parts often require new production processes, and new equipment, dies, fixtures or tooling, and associated extra changeovers; these burden already-loaded shop-floor activity.

   4. Congested Floor Space, Shelving, and Storage Racks. As with parts in general, new parts need homes, however temporary, and add clutter to floors, racks, shelving, and stores; over time, multiplying service parts and dead stock (obsolete but not yet retired parts) further cramp already congested storage areas.

   5. Overburdened Material Handling. If the company accepts material handling as a given, added parts tax an already burdened transportation system, as each new part requires its share of handling in the form of receiving, counting, inspecting, storing, retrieving, and otherwise moving it.

   6. Ballooning IT Input and IT Maintenance. Each part that enters or leaves the system must be individually logged in and maintained; pressure to keep data systems up to date can be staggering.

   7. Mushrooming Control Points. Literally hundreds of paper, computer, and other transactions across all departments (known as control points) support each new product and its new parts—drawings, catalogs, cost estimates, supplier searches, purchase orders, faxes, invoices, receipts, tracking checklists, inspection sheets, etc.

8. Loss of Opportunity. The resources needed to support runaway parts proliferation are astounding. They rob the company of assets it could otherwise use to develop and grow (Figure 2.4).

Look at the eight runaways on the list. Taken as a series of single events, it is not hard to justify the need for each of them in the successful running of the business. But look again. In the actuality of profit and loss, they are each non-value-adding in the same sense as the previous set of operational wastes are. That is, while many of the by-products of new product expansion do not create problems for the organization in and of themselves, each one becomes a significant problem when it is recurrent—when it happens in sufficient multiples. Then it becomes a runaway and a genuine source of complication and unwarranted cost.

In multiples, these by-products provoke day-to-day entanglements that, in turn, can cause the company to tilt into overload and go into organizational gridlock. If we are to head this off, or be empowered to make a midcourse correction, we must first recognize these by-products for what they are—waste!

Figure 2.4. **The Eight Runaway By-Products of New Product Expansion**

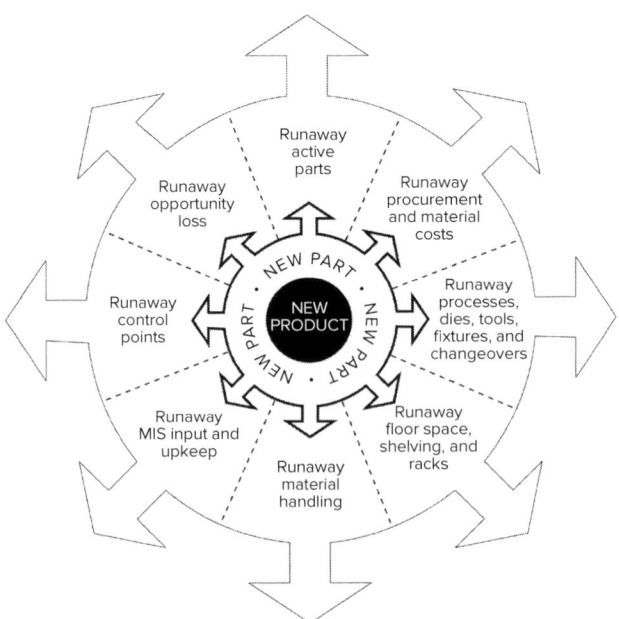

Rethinking the 95:5 Ratio—Death by 10,000 Cuts

In the decades preceding the introduction of JIT (precursor to lean), inventory was erroneously presumed to be an asset at best or a necessary evil at worst; as a result, the wastes it created remained invisible for a very long time. Similarly, many companies continue to consider the *by-products of new products* necessary and required—the price paid for market share.

In fact, these by-products are monsters, hidden deep in the infrastructure of the company and, like excess inventory, they drain the life force of the enterprise. Without a process such as VEP, they will never become visible and therefore never be confronted and addressed. Having said that, let's take another look at the 95:5 ratio we previously discussed in light of new product expansion. There is a valuable insight there that should not be overlooked.

We begin by stating the obvious. The sale of a product and subsequent revenues are critical to a manufacturing company—the doorway to its profits. The development of new products initiates that process; and so we place new products on the value-adding (VA) side of the ratio (5%). The 95% or non-value-adding side is comprised of those wasteful activities that occur in the process of getting new products introduced. These are the equivalents of the *seven deadly wastes of production* (making defects, delays, overproducing, over-processing, motion, etc.). Until you are aware of them, these can often be as invisible as non-value-adding activities in production were—just "part of the way business gets done around here" (Figure 2.5).

Figure 2.5. **The Seven Deadly Wastes of Production: Equivalents in New Product Development**

7 DEADLY WASTES OF PRODUCTION	EQUIVALENTS IN NEW PRODUCT DEVELOPMENT
Making Defects	Making mistakes in design
Delays	Waiting for specs, waiting for approvals
Over-processing	Reworking the design, conducting multiple tests
Overproducing	Creating multiple prototypes
Motion	Searching for information, missing drawings, etc.
Making Inventory	Adding new parts
Material Handling	Circulating designs, prototypes, test data results
Missed Opportunities	Loaded down with current design work, the company cannot move on to new developmental opportunities

In reality, these wastes are concealed, embedded in the 5% or value-adding side of the ratio. If we explode them out and into their true 95% impact, we begin to recognize their enormous non-value-adding drain on the organization (Figure 2.6).

Figure 2.6. **95:5 Ratio in Runaway By-Products Hidden in New Product Activity**

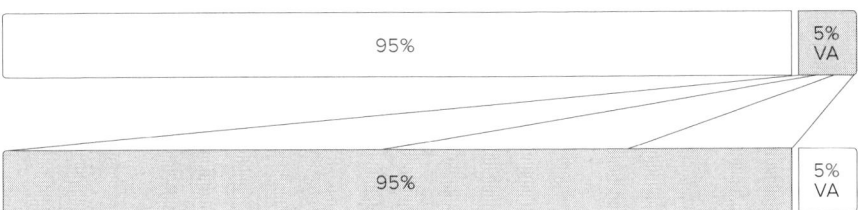

The first challenge was to eliminate the seven deadly wastes from the flow of production. Today, the challenge is to remove the silt accumulating on the river bottom from the runaway by-products—unneeded drawings, processes, dies, tools, racks, changeovers, etc. These are making the river of production increasingly shallow and congested and threaten to choke off the flow. But they are, like the silt itself, mere symptoms of a deeper blockage, the true cause of the decelerating production flow. The true cause lies under the silt—those myriad small, tangible objects lodged in the riverbed. We call those objects *unwarranted parts—ineffective variety*. Their removal and prevention is the focus of VEP (Figure 2.7).

Figure 2.7. **Blocks in the Production Flow: Rocks Revisited**

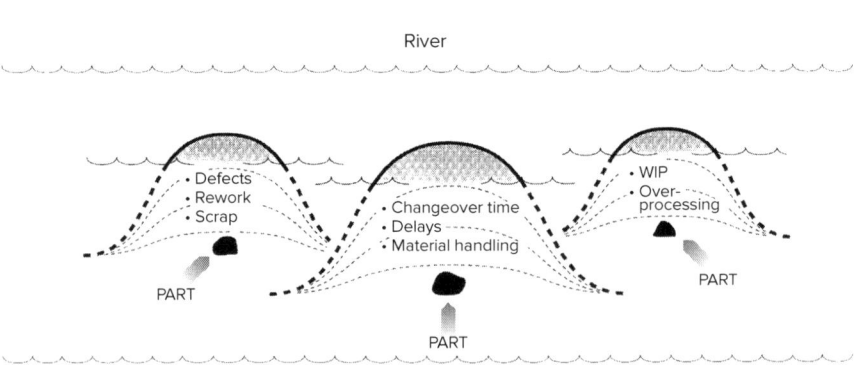

Many manufacturing companies release an avalanche of new parts into their systems every year. Like the famous Chinese proverb describing a slow and initially painless death by ten thousand cuts, each part exerts its tiny stress on the enterprise. Eventually, the company tips over into chronic complexity and organizational congestion. Let's look a little closer at this phenomenon.

New England Farmhouse Effect

Profit-making is a constant trade-off between revenue and cost. But many businesses nowadays buy increased sales through new products and make no attempt to control the runaway by-products. As a result, any sales increase is bought at an inflated cost rate. Too few new products are commercialized as a result of a carefully defined and executed product introduction strategy. In many instances, an organization's new product approach resembles how farmhouses in New England often get built—expand as needed in any shape and direction (Figure 2.8).

Figure 2.8. **New England Farmhouse Effect to New Product Introduction**

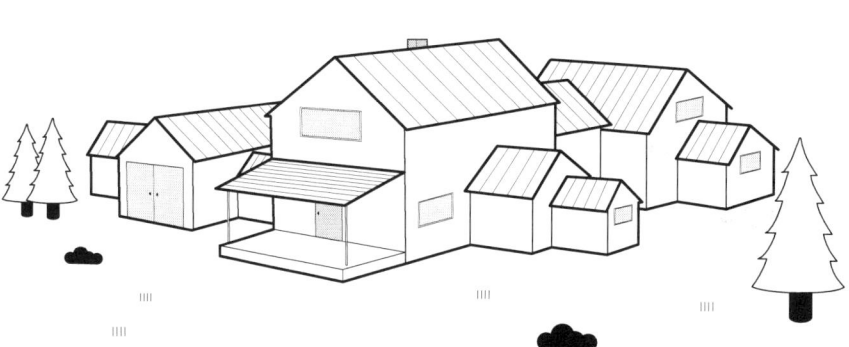

When new products are introduced, sales appear to respond and grow. The number of parts, however, is escalating at a rate parallel to or greater than those sales, and a dangerous Y-shaped curve begins to emerge as the rate-of-parts increase overtakes the rate-of-revenue increase (Figure 2.9). Profits begin to erode and the company wonders why.

Even when a company achieves genuine market-driven product diversity, there is no guarantee that profit will improve. In fact, the reverse often happens. Product variety increases—but output grows only slightly, stagnates or even declines. Concurrently, costs escalate, sales slow, and profits erode.

Figure 2.9. **The Y-Type Trajectory: Profile of a Company in Trouble**

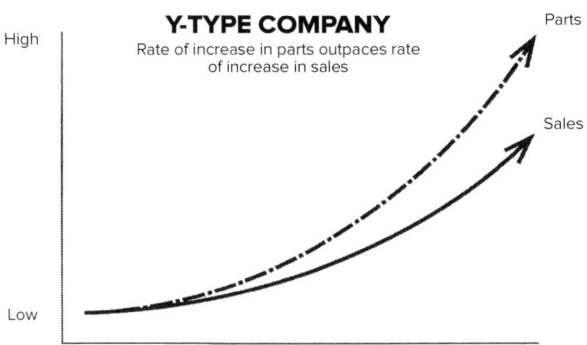

In the face of this, the company may attempt to regroup and bolster flagging sales by developing still more new products in the desperate attempt to rectify the cost curve. Gross sales may go up in response, but profits usually do not recover. Meanwhile, sales on existing products can shrink. Marketing looks for new product opportunities. Design and Product Engineering try to speed up new product development. Production introduces more high-performance equipment and acquires more storage space and purchases more forklifts to speed up material handling. Production control labors through the introduction of new computers and next-generation software. Neither production nor sales can keep up with the burden of increased product variety. Generating more costs than revenues, the company sinks deeper into financial quicksand with each new product offering. The entire company engages in a futile attempt to build a box big enough to hold its burgeoning product line.

Companies caught in this Y-type trajectory pay for their products twice: once as part of the official product introduction budget, and again in the cost of managing the complexity added to their systems as a result of introducing new products.

PUI: When Variety Is Negative

In retrospect, Parts Unlimited Inc. became a victim of its own growth. Product variety—and the ancillary activities it triggers—are out of control.

Although company leaders failed to recognize the implications at the time, every new product that got introduced served to fold in yet another microlayer of cost and complication. Like many companies, PUI took decades to blunder its way into trouble.

What went wrong at PUI was not its commitment to maximizing customer selection and seizing market share. Product variety is not the enemy. What went wrong was the company's lack of an equal and parallel effort to regulate and control the negative side effects of that selection. Negative variety—the sum effect of the runaway by-products of new products—got the upper hand.

The fact is, the central premise of PUI's product approach is both erroneous and dangerous. In order to continue to meet and exceed the need of its customers for new products and retain market share, the company believed that its parts inventory had to increase. Negative variety, thought PUI managers, was inevitable.

VEP: The Alternative

Negative variety is not only *not* inevitable—it is avoidable. But in order to avoid it, organizations must systematically head off the negative aftermath of new product introduction—the downside of product expansion—*before* it takes root. Managers need an approach that will help control and then reduce their parts inventory and dismantle existing complexity from the inside. They need to do this even while they initiate new practices and policies that prevent negative variety from recurring. Attempts to eliminate the problem by stemming the flow of new products does *not* solve the problem—but instead are likely to sink the enterprise. The only solution is to understand and eliminate the true causes of the problem and stop the downward spin.

Variety Effectiveness Process (VEP) provides a solution. VEP enables companies to step back and dismantle the layers of complication that cover and choke the organizational infrastructure. Through VEP, unneeded products, parts, production processes, and control points are identified and minimized.

> *VEP is a systematic team-based methodology directed at maintaining or expanding customer selection while reducing negative variety in parts, processes, and control points and preventing their future recurrence. Its goal is to lower costs dramatically and de-complicate systems while maximizing a company's ability to respond to the demands of the market.*

Effective implementations of VEP can result in reductions of 25% to 40% in parts count, 15% to 40% in production processes, and as much as 60% in control points (those transactions aimed at procuring, receiving, inspecting, storing, counting, and retrieving parts and products).

The Rewards

Such reductions create deep and far-reaching benefits for the life and flow of the enterprise. When effectively implemented, VEP does not simply reduce the number of parts, processes, and control points. It frees the system from the inside to achieve new levels of health, flexibility, and vigor. From single-product industries like bricks to multi-product ones specializing in household appliances, furniture, or control equipment, companies that adopt the VEP perspective can experience impressive rewards:

1. Reductions in Total Active Parts. Through VEP, parts are eliminated or combined, and subassembly levels simplified. As a result, bills of material are streamlined and flattened, requiring fewer layers because products are developed based on the principles of effective variety. Reduced parts count, by association, triggers all the other rewards that follow.

2. Fewer Production Processes, Special Equipment, Dies, Tooling, and Machine Changeovers. Eliminating even a single part often triggers parallel reductions in production processes, dies, tooling, and the number of machine changeovers. The cumulative impact is immense.

3. Reduced Storage Space and Less Material Handling. Reductions in products and parts automatically erase the need for their handling. The need for shelving, racks, and other storage shrinks along with the space formerly required for associated production processes, conveyors, material handling, fixtures, equipment, and the office and support areas that were needed to control and maintain such units. Literally miles of square footage can get freed up.

4. Accelerated, Complete, and Accurate Parts Information Retrieval. Because intelligent design is closely linked to the capability of the parts database, VEP works to ensure that a company's parts classification system supports strategic design decision making.

5. Strengthened Functional Alignment between Engineering, Operations, and Sales and Marketing. VEP is a team-based approach that utilizes information and insight from all the players involved in product design, procurement, and manufacture. This multi-functional approach breaks down the barriers between functions as people focus cooperatively on the task at hand.

6. Reduced Product Introduction Lead Time. With efficient database and design practices in place, it takes much less time to go through the development process. Marketing becomes fully cognizant of the implications of product and parts variety, as does Engineering. In addition, Design and Product Engineering has easy access to relevant, complete, and accurate information for making sound decisions about new parts.

7. Reduced Levels of Paperwork and Other Control Transactions. Since VEP sees the part as the ultimate cause, the cumulative addition of a single part number can, over time, trigger a vast range of activity that would never exist without it. As parts are eliminated, procurement and other supporting activity are simultaneously and dramatically reduced. In addition, the VEP method provides for specific efforts to streamline and reduce remaining control points, independent of parts reduction procedures.

8. Upgraded Policies and Practices To Prevent Future Proliferation. VEP seeks the root causes of product and parts proliferation. In many cases, these causes are concealed in seemingly blameless policy directives and day-to-day practices, such as those related to purchasing, equipment, and design. When a company's corporate policies and practices are revised to align with VEP principles, many causative factors disappear.

9. Streamlined Procedures for Engineering Change Notices (ECN). When engineering changes must happen, they are moved through a simplified procedure governed by now-familiar principles and practices that prevent unneeded variety from entering the system (Figure 2.10).

Companies adopt the VEP approach because it is quite simply too expensive for them not to. VEP is especially relevant in any company in which product diversification is a strategic issue and proliferating costs, a persistent concern. Companies in a range of industries—from electronics to home furnishings, from pharmaceuticals to utilities—all can benefit from VEP. From the machine tool industry to shoes and clocks, companies can gain a significant competitive advantage by applying the VEP Methodology.

As you will read in subsequent chapters, with VEP's emphasis on cross-functional analysis and aligned change, it cuts through the issues that cloak negative variety and it illuminates its causes and solutions. The

VEP process represents a new way of designing and developing products that positively impacts not just those products but all the activities and transactions that support them as they move through the workplace on the way to end-users.

Figure 2.10. **VEP Outcomes: A Chain of Rewards**

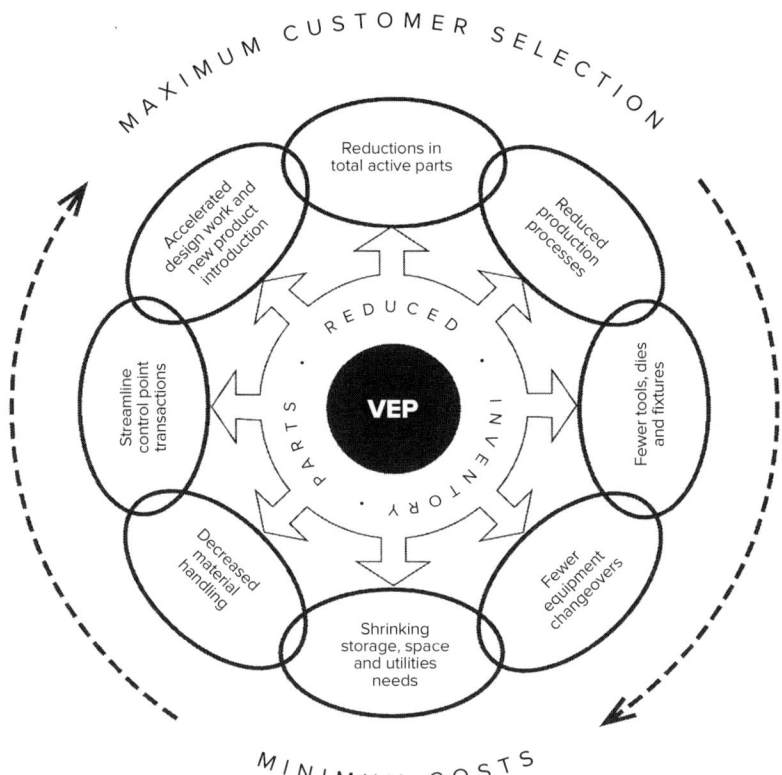

What Is Variety Effectiveness?

Variety effectiveness is a new way of looking at product expansion. No one can deny the criticality of a company's introducing new variety into its product lines. But this variety must keep cost and complication to an absolute minimum. When it does, the variety that results is said to be effective. By the same token, negative variety does not appreciably expand selection but adds cost that the company may or may not be able to pass on to the customer. In short, variety effectiveness refers to the extent to which:

- variety within new products contributes to profit
- variety within existing products is as much as possible customer-driven, with any associated negative variety—including control points—minimized
- unwarranted variety (which is always negative) is actively and systematically stemmed or eliminated

The X-Type Company

The job of the VEP Methodology is to assist companies in cleaning up the negative variety from the past—and ensuring that *all* new variety is as positive as possible. With negative elements removed or minimized, the true profit potential of a company's products is achievable. When this happens, the *Y-shaped curve* previously described in Figure 2.9 makes a midcourse correction, with sales steadily rising even as parts take a downward turn. The result is the profile of the X-type company (Figure 2.11).

Figure 2.11. **The X-Type Curve: Profile of a Company Succeeding**

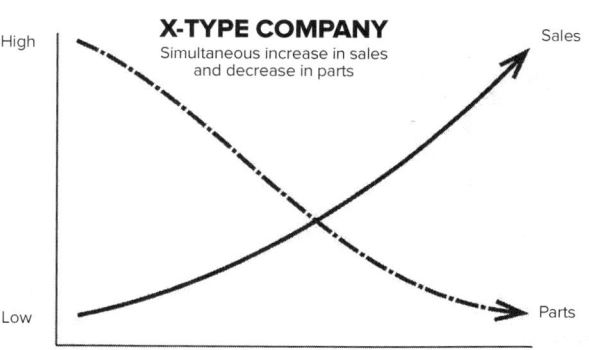

In the X-type company, contrary to the parallel escalation of sales and costs, the continued attention to parts reduction allows sales to rise as costs drop, a far more successful outcome. This does not happen by accident. The X-type profile emerges over time as a result of a well-defined, intentional, carefully implemented strategy of variety effectiveness. Products are made for profit from the get-go, and levels of complexity are manageable. They are designed to balance the quality and speed required for low-cost and high-impact production with the richness and ingenuity that are needed to offer truly viable new products and to expand markets. Penalties for creativity are kept to an absolute minimum. People know what levels of variety are warranted for specific levels of output and return. They can spot negative variety at a glance.

In the X-type company, products increase in response to market demand—but the total number of new parts and processes increases at a lesser rate. The company's approach ensures no deterioration in profitability throughout the life of its products, even after the market peaks and begins its decline. The business has already found satisfactory answers—VEP answers—to key questions such as these:

- Which market niche are the characteristics of each product in a product group designed to satisfy?
- How many parts are now used to provide the characteristics of each product—and how many are optimal?
- How many production processes are now used to produce each product at the required level—and how many are optimal?
- How many control points are now used to ensure quality, delivery, and customer satisfaction—and how many are optimal?
- What specific corporate policies and business practices are in place to support effective variety and enhance market share and profit? What new policies and business practices need to be developed?

In such companies, smart simple design is a strategic commitment and a living reality.

The goal of VEP is to help mature companies shift from a Y-type to an X-type profile—and to guide new organizations toward practices that create X-type results and avoid Y-type triggers.

VEP'S Multidimensional Approach To Effective Variety

How does VEP help companies avoid becoming Y-type organizations or make the transition from Y to X? VEP meets this challenge by linking five powerful dimensions: 1) a comprehensive view of cause, 2) an in-depth analysis of complexity, 3) an insight-rich, team-based approach to improvement, 4) the unraveling of complexity through engineering-based tools, and 5) a prevention-based mindset. These are summarized below and then discussed in detail in subsequent chapters (Figure 2.12).

Figure 2.12. **The Five Power Points of the VEP Approach**

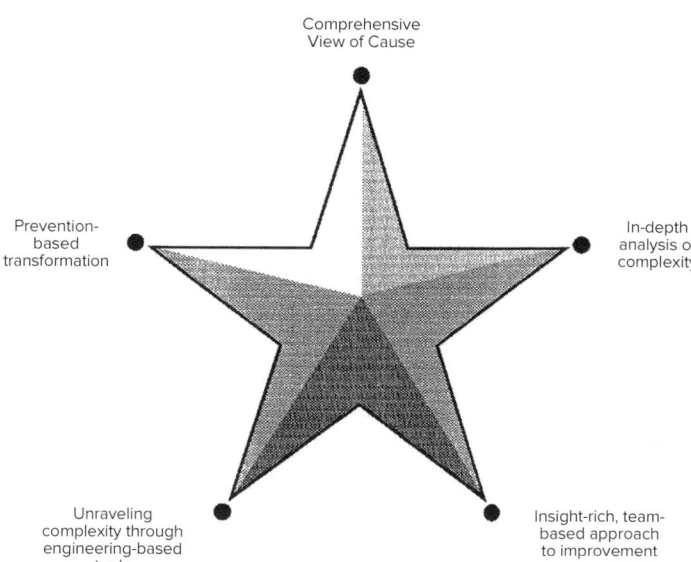

1. Comprehensive View of Cause

Companies can pick from a wide choice of improvement methods that attack waste. But they should take care. Few target the full array of causes that trigger negative variety as discussed in this book. So while these methods will produce some benefit, they are not enough.

Let's look at some of the many techniques that support improved product development. For example, value analysis/value engineering, quality function deployment, and other market-in techniques can successfully reduce parts costs and increase product value on a unit-by-unit, customer-by-customer basis. Costs across the board as well as organizational complexity, however, can still increase. *Standardization* is another tightly focused cost-cutting approach; it focuses on normalizing products by communizing components and subassemblies. Still, customer selection is often sacrificed.

VEP adopts a wider view. While maintaining a tight customer focus, VEP dismantles the internal obstructions caused indirectly or directly by the introduction of product variety. To succeed in this, a company must assess formal and informal policies within each organizational function as potential triggers of the runaway by-products of negative variety. Broadly speaking, these functional areas are:

Marketing and Sales. In a sometimes overzealous search for differentiation between products, certain marketing practices—including product nomenclature—trigger layers of variety that are unwanted by the customer and unmanageable by the company.

Product Development. There are many ways that unwarranted variations in products and parts creep into the design process—high designer turnover, lack of a unifying development strategy, designer preferences, changing technology, the absence of sound information to support design decision making, and shifting design criteria, to name a few.

IT/Data Systems. A company's computer database and classification systems often make it difficult for Marketing and Engineering to make decisions in favor of effective variety because the information housed in these systems is inaccurate, incomplete, obsolete, difficult to retrieve or categorize and understand.

Accounting. Few accounting approaches recognize unwarranted variations as the framework within which costs and complications escalate. Traditional accounting procedures mask these costs, making them the hidden price of doing business. Although concealed, these costs tighten their grip on the company and continue to drain off its resources.

Operations. Flawed operational assumptions, such as the belief that large lot production is required, can trigger unneeded variation. This problem is further exacerbated when the company feels compelled to purchase equipment capable of running high volumes and considers high inventory an asset.

Chapter Four is devoted to discussing these and other trigger areas in detail.

2. In-Depth Analysis of Complexity

De-complicating enmeshed internal systems is no easy task, particularly in mature organizations where negative variety has taken root over a number of decades. The only choice is to start at the perimeter and work inward.

VEP does this through its central approach, the *3-View Analysis*. This analysis focuses equal attention on: 1) the marketing factors that trigger the expansion process, 2) the structure of the products themselves, and 3) variation found within each parts type in the company's parts inventory. The objective of this three-way approach is to identify opportunities to optimize positive variety, reduce negative variety, understand the causes of negative variety, and implement changes in the organization in order to remove those causes and ensure that future negative variety is minimized. Each of the three views represents an independent framework of inquiry— in essence, a series of questions that, when answered, shows us where the greatest opportunities for reduction lie.

In another sector of in-depth analysis, VEP shines its light directly on the unwarranted proliferation of production processes and control points. Further reductions are then made.

3. Insight-Rich, Team-Based Approach to Improvement

Organizational complexity is not triggered by single causes but rather by a network of highly diverse, interlocking variables. Many hands are needed to unravel it and get the company back on track.

In doing this, a company can choose one of two VEP implementation options. It can adopt a narrow focus (called the *discrete* approach) that uses a crack task force of internal experts to analyze and improve a specific product line. Or it can adopt a comprehensive application (called the *deep-dive* approach), bringing the organization together on cross-functional teams to reduce negative variety company-wide.

In either scenario, VEP recognizes people as the resource for insight and creativity, and encourages a team-based process that capitalizes on the knowledge and experience of a wide assortment of company employees. VEP provides the logic, tools, and structure for them to succeed.

4. Unraveling Complexity through Engineering-Based Tools.

The job of de-complicating the infrastructure of a company can be daunting, particularly when, for decades, product expansion has not been guided by a unified development approach that builds on the principles of variety effectiveness. VEP turns to an engineering tool chest for six techniques to organize and analyze product- and parts-related data into the meaningful format required to solve the problem through implementable improvements. VEP teams adapt these six analytical tools, (called the Six VATs) to the focus of their individual inquiry—whether that focus is marketing, product structure, parts types, control points, production processes or corporate policies and business practices. (See Chapter Eight for a detailed discussion of the Six VATs.)

5. Prevention-Based Transformation

The work of a VEP implementation is not considered complete just because parts inventories shrink by 30%, or because 25% of all production processes are eliminated, or because control points are reduced by half. VEP's job is not done until the roots of negative variety have been dug out of the enterprise and replaced with practices and policies that will prevent their recurrence. To do this, VEP also knows that an array of stakeholders (all departments and suppliers, plus the customer) must be satisfied and their interests respected. A balance is struck between competing organizational forces.

VEP is about a prevention-based transformation that looks to the long term for results. What is your parts count one year after VEP victory was declared? Two years after? Ten years after? Has the company stayed the course, maintained its optimal parts count in the face of even greater competitive threats? Has it remained a *least-cost sum* company? VEP sets up a dynamic infrastructure that ensures that the answers to all these questions are affirmative and negative variety cannot regain a foothold in the organization.

VEP: Not Magic—Work!

The Variety Effectiveness Process is far more than the latest entry in world-class improvement methodologies. The issue of the effectiveness of the variety a company offers, whatever the market, has been long *unrecognized* as a success factor. Yet, for many companies, this is where a true competitive advantage lies. De-complicating systems that have evolved over decades is not an easy task; inevitably some painful restructuring must occur. But for the organization that recognizes the potential rewards, such an undertaking will be welcomed and diligently implemented. The stakes are high and the perils of failing great. All-around business excellence is the only goal worth seeking. VEP is a core element of that quest.

On The Horns Of The Dilemma

Before we examine the details of the VEP method (the *how*), let's find out more about VEP's core principles, how it views cost, where policy issues obscure both the problem and its solution, and revising the design mindset.

CHAPTER 2

CHAPTER
2

CHAPTER 3
True Cost: Product Proliferation And The Bottom Line

Does the cost system you are now using take into consideration the true cost of your products or the true cost of diversifying your product lines?

Creating a world-class enterprise requires a fundamental change in the way a company approaches the design, production, and distribution of its products and services. To achieve effective variety, VEP can play a key role by helping a company track and reduce the level of its parts inventories and internal complexity.

VEP offers two tools, discussed later in this chapter, that drive and measure the company's progress toward greater profitability: The VEP Parts Index and The VEP Tri-Cost/True-Cost Model. As a result of applying these tools, a more valid, immediate, and practical tracking approach emerges—one that helps the company valuate and understand the complexity of its internal systems and the true cost of making product.

These two VEP mechanisms offer a perspective on cost that is at variance with the more traditional approach known as GAAP *(generally accepted accounting principles)*. The GAAP approach gauges a company's financial progress almost exclusively on outputs and past performance. With its focus on measures such as labor, material, and overhead, GAAP practices can actually interfere with improvement efforts and often run counter to world-class values and objectives. Created nearly a century ago at a time when accumulating parts inventories was not a key corporate concern, GAAP was not designed to track negative variety and the unwarranted proliferation of parts. The reverse was more often the case: Because GAAP could not and did not assess burgeoning parts counts, it often provoked further parts accumulation and masked the direct and the indirect consequences of negative variety.

Let's visit Parts Unlimited Inc. (PUI) again and find out more about its accounting approach and the problems facing Tom Vargas and his company.

A Case In Point: Accounting Gone Awry

PUI now generates revenues of about $25 million a year from the sale of mechanical and electronic process controls. Selling to both distributors and end-users, its products can be found in applications ranging from food processing to oil and natural gas exploration.

The Vargas family has owned PUI since its founding during the Depression. Over the years, the workforce has grown steadily from a 15-person garage shop operation to more than 200 people currently employed. Today, that work force is composed of 40% direct manufacturing employees and 60% support personnel. Ten years ago, the work force was 50% direct labor and 50% support. Twenty years ago, the work force was 55% direct labor and 45% support.

PUI operates a single manufacturing plant and relies largely on hand assembly to build most of the 100,000 standard end-item configurations that comprised its 50 core product series. Supporting processes include: welding, cutting, filling, machining, wave soldering, computer numerical control machining, testing and setting, and component burn-in.

As we learned in the last chapter, PUI management has become alarmed at the company's ballooning inventory figures. Despite a relatively low growth in revenue (2.3% a year) over the past five years, inventory has risen from $4 million to $5.6 million last year. In the same period, PUI's active parts count grew from 9,454 to 13,156, an average increase of 6.8% per year. A variety of reasons were offered to top management to explain this increase:
- New products require new parts.
- To get more market share, PUI must offer whatever the customer wants (be everything to everybody in the market—whatever it takes).
- Customers want things faster so PUI has to have more inventory.

In addition, lead times from suppliers are horrendous; parts shortages threaten to—and often do—close down lines.

As we heard, Tom Vargas is unwilling to accept these conditions, particularly in view of the current economy and the company's declining profitability. With another year of anticipated low growth on the horizon and a prediction by the materials group that the inventory figure will grow to $6 million by year end, Vargas is convinced the time for action is now.

He is also concerned because he knows that inventory costs are just the tip of the iceberg. Recently he attended a seminar about VEP/Variety Effectiveness Process that discussed hidden inventory costs that arise from producing, inspecting, and storing massive quantities of parts—as well as costs related to planning, organizing, and controlling them. At PUI, these consequences seem to multiply by themselves and have quickly outpaced any increases in sales. Vargas recently heard his firm described as a typical Y-type company, a direct reference to this parallel rise in sales and parts inventory; he knew it was not a compliment to his managerial skills.

At PUI, slow and moderate product growth is coupled with rising inventory levels. But that is not all that is increasing. Even the company's modest product growth has triggered:

- More setups and changeovers in production processes
- New tooling and dies
- Additional floor space for production, inspection, and warehousing
- New production processes and investment in equipment to support them
- Additional suppliers and purchasing costs
- New packaging and shipping materials
- Increased design time and expenses
- Mounting quality-control costs
- Increased accumulation of outdated stock
- More time for inventory planning and control
- Additional documentation requirements such as new drawings and procedures
- Rising opportunity costs (money committed to inventory and its support that cannot be used for other purposes)

CFO Helen Leary shares Tom Vargas's concern about the burgeoning inventory values. But she also has severe reservations about the company's ability to measure the actual costs associated with all these parts, and with related activities such as purchasing, paperwork, and drawings. The current computer package at PUI incorporates a costing module that measures product cost based on material costs and direct labor, with an additional amount factored in for company overhead. This overhead is calculated for each product based on the amount of labor in the product—but does not take into consideration the amount of inventory, inventory handling, or control tasks related to the product.

For these reasons, Leary feels the current method of costing is not conducive to a proper understanding of the real cost of a product. For example, she knows of one popular Series 11 product that has few parts in it but requires more direct labor than the comparable, slower-moving Series 8 product that has almost twice the parts. Under the existing labor-burdened system, the Series 11 product absorbs more of the factory overhead and has a higher total cost—even though Leary is convinced that, because it has fewer parts, it actually costs the company less, overall, than the Series 8 product (Figure 3.1). Leary and her staff see the burden of the mushrooming parts on the company; but it is practically impossible to validate what they see using the principles and techniques available to them in their traditional cost-accounting package.

Helen Leary's concern with what is happening at PUI is an important one. The dilemma she faces was created long before she came to the company. Let's look at the origins of GAAP.

Figure 3.1. **True Total Cost: Which Product Costs More?**
(Comparison of Series 11 Series 8 Products)

COST CONTRIBUTOR	SERIES 11 PRODUCT	SERIES 8 PRODUCT
Direct Labor	21 minutes	10 minutes
Material Content	$20.50	$21.75
Total Parts	29	46
Production Process Steps	6	16
Tools & Dies	1	4
Fixtures	3	7
Control Points • Drawings • Transactions, Paperwork, Inspection Points	33 8	51 12

The Traditional Cost Approach: History And Logic Of GAAP

In the early days of the industrial revolution, *cash flow* was the central barometer of fiscal viability. For all intents and purposes, the tin cash box was the only available cost-management tool. It worked like this:

1. If there is enough money in the cash box at the beginning of the day, expenditures are made. If not, they have to wait.
2. If money is left at month's end, the business has turned a profit and the enterprise continues. If the cash box is empty, the enterprise folds.

As the U.S. economy grew, companies grew with it; manufacturing sites multiplied, more and differing products were offered; distribution systems began to operate on a nationwide basis. Industrial growth became increasingly dynamic and was no longer manageable in simple cash box terms. Cost accounting was born.

By the 1930s, basic accounting concepts and practices were formalized into *generally accepted accounting principles* (GAAP). Many of these same cost-accounting practices are in use today. At their foundation is the following definition of product profit—which is also of special interest to the VEP discussion:

PROFIT = PRICE − COST

Profit, according to GAAP, is the difference between product cost and product price—and continues to be. Here is how GAAP (or traditional accounting) calculates cost, specifically product cost:

PRODUCT COST = MATERIAL COSTS + LABOR COSTS + OVERHEAD COSTS

We can better understand how a traditional cost accounting approach triggers complexity by looking at the definition of each of these cost terms: material, labor, and overhead costs.

Material Costs: A direct cost that includes the purchase price of each part and/or raw material used to produce an item, as listed on the bill of material (BOM).

Labor Costs: Also a category of direct cost, labor represents a combination of manpower level and the level of mechanization and automation required to convert the parts list (BOM) into the desired level of product.

Overhead Costs: Those costs that are not directly assignable to a given product and are, therefore, allocated across all products on a formulaic basis. Overhead or indirect costs include depreciation on equipment, heat, light, power, taxes, research and development, maintenance—as well as salaries and wages for operations-support personnel and fringe benefits for both direct labor and operations-support personnel.

Based on these, the GAAP product/cost formula assigns a standard cost to each manufactured product. Standard cost represents the sum of the costs of each individual part in a product and each labor step in a production process—as well as associated overhead costs assigned to the product, based on some pre-determined formula. In this sense, the standard cost approach serves as the framework against which actual costs are assigned, entered into the cost-accounting system, and then analyzed for variance. Variance data are then linked to the level of each department's productivity and, if deficient, to production problems.

Straightforward as this procedure may appear, it poses certain challenges in trying to track proliferating complexity in an organization.

From Tracking the Past to Tracking Complexity

Standard cost accounting is a set of assumptions intended to help company managers evaluate organizational strengths and weaknesses to determine if the enterprise is on or off the profit track. Linked to these assumptions is a complex system of performance indicators, measures, and techniques for pricing products and assessing operational efficiency. When GAAP techniques were introduced decades ago, an information conduit was formed relative to the operational and financial performance of the enterprise. Companies used this information for decision making relative to a vast array of strategic, tactical, and day-to-day needs.

In the late 1970s, the way companies did business began to change. The accounting system developed 60 to 100 years earlier to track that business, however, did not change. A value divergence developed. To understand the depth of this discrepancy, remember that any accounting approach—traditional or otherwise—is a measurement and reward system at its foundation, a system that defines which performance outputs are valuable and which are extraneous. Given this, the choice of accounting system is as significant to the future of an enterprise as the profit and loss it measures.

At their best, performance measures align with the company's vision, pointing the way to the strategic and tactical decisions needed to achieve that vision. At their worst, ill-conceived performance measures deflect an organization from its path of growth and achievement, causing it to founder, even if measures signal otherwise. That is the nature of measurement—and part of the problem with GAAP related to VEP.

Variety effectiveness is reached when two conditions are met:

- The company has achieved optimal levels of parts inventory to support its market growth
- Company systems are sufficiently de-complicated to require *least-sum* resources.

As we will see later in the chapter, this understanding can trigger practices that lead to decreases in unwarranted parts and in other direct and indirect wastes.

The problem in traditional managerial accounting is that a product's cost is based almost exclusively on the three factors defined above: materials, labor, and overhead. As illustrated in Figure 3.2 and the discussion that follows, the GAAP approach causes more parts to be purchased or made; this, in turn, leads to more (not less) internal complexity and higher (not lower) parts inventories. From there, complications adhering to new parts skyrocket. Let's take a close look.

Figure 3.2. **GAAP vs. VEP**

GAAP Approach

- Tracks the past
- Assigns disproportionate product overhead
- Focuses on least-cost parts/piece price
- Triggers multiple suppliers
- Triggers large lots and machine over-utilization
- Promotes price mark-up to attain acceptable profit

VEP Approach

- Tracks complexity and needless variation
- Tracks total parts count
- Focuses on true cost (F-, V-, and C-Costs)
- Promotes single-source suppliers and global pricing
- Fosters price-targeting and design-to-price practices
- Encourages small lot production

Does your accounting approach focus on the bottom line or micro issues—on total cost or the cost of individual products?

Flaws In Traditional Cost Accounting Assumptions

The overall goal of cost accounting is two-fold: first, to ascertain how much it costs the enterprise to provide products and services to its customers; second, to determine how much the company gains in profit from these endeavors. Any flaws in the set of assumptions that govern these two goals can have serious repercussions for the enterprise, no matter the approach. GAAP's flawed assumptions relate particularly to cost and price, the two elements in the traditional definition of profit.

Flawed Assumptions about Cost

Material Costs. A hundred years ago, material costs accounted for about 10% to 15% of total manufacturing costs in the ratio among labor, material, and overhead. Due in part to decreases in the two other formula elements (labor and overhead)—as well as a shift from labor-intensive work to work done by machines—the level of material cost has grown in recent times to 50% or more of total cost. In addition, material costs typically represent the price for which materials are delivered at the facility. But the GAAP formula does not adjust, for example, for procurement or carrying costs.

Labor Costs. When accounting techniques were developed in the early days of the industrial revolution, labor costs accounted for 75% to 95% of total cost, the largest share by far; material costs came next, followed by overhead costs. But cost patterns have changed significantly in the past 50 years due to new product and new production technologies. The result is a marked reduction in the amount of labor required to manufacture a product.

By 1990, the labor content of the average product manufactured in the United States, for example, was 7% of the total cost. In isolated cases, where commitment to JIT (and then lean) and other world-class techniques was high, labor costs dropped even lower. In companies where the labor content formula remains unchanged, the accounting department continues to expend significant time and energy to track and valuate labor elements that have become increasingly less important to the profit equation.

In addition, while standard cost accounting makes it fairly easy to calculate the labor rates (wages paid by category) and time requirements for the operations needed to complete a product, other factors that figure prominently in driving up labor costs are more difficult to reckon. These include: the size of production runs, number of different runs through the same machine or department, equipment changeovers, operator experience plus the detrimental effects of high employee turnovers, fluctuations in production output, and the precise link between a specific product and the processes (production and non-production) required to realize it. As a subset of difficult-to-reckon factors associated with product-linked processes, make sure to also include any searching for or moving of parts, tools, fixtures, paperwork, etc.

Overhead Costs. According to GAAP cost-accounting principles, overhead costs are not attributable to specific products; as a result, they are allocated to *all* products. Seventy years ago when this practice was codified, this did not pose a problem because the overhead costs of a product were negligible compared to its labor content. There was little need to track them in great detail.

Decades later, the differences between direct/indirect and fixed/variable costs are far less distinct. The boundaries between them are blurred. The overhead portion of product cost can account for as much as 35% to 50% of total cost—and, in some cases, can escalate to 75%. The net effect is that many products are assigned disproportionately large or small overhead share. Since overhead now represents a large measure of total product cost, this inequitable formula for analyzing and assigning has serious repercussions for the enterprise.

Some companies have adjusted their standard allocations to correct for these outdated cost assumptions. Others have gone over to a system of activity-based costing (described later in the chapter). But in many companies, traditional cost-accounting practices remain unchanged. If a company committed to the path of world-class values and results continues to use these, it may get a distorted and misleading picture of its present performance, challenges, and improvement options. This is perhaps the greatest danger.

We now turn to the price component of the profit formula and some erroneous GAAP assumptions associated with it.

Flawed Assumptions about Price

Shift from Margins to Market Share. Traditional cost analysis is geared to calculate product price based on acceptable margins. Over the past several decades, however, there has been a shift in corporate strategy, away from customary pricing practices to one that assigns price based on market share and long-term company viability. The emerging importance of capturing new market share has fueled a rush among manufacturers to introduce a continuous stream of new products at a price the *consumer* finds acceptable.

Cost-Driven vs. Market-Driven Pricing. In the traditional accounting approach to pricing, a company arrives at a suitable selling price by putting an acceptable profit margin on top of its costs. Product price is a function of cost differential and markup; over the decades, many a company has improved its bottom line simply by raising the price tag. In this scenario, price becomes a matter of what the consumer will bear.

But that day has passed. The economy (both domestic and global) has entered a period of disinflation in which surging demand fuels lower, not higher, prices. Low-ball or target pricing is now king. These changes in pricing patterns are positive ones and do not diminish the role of cost as a profit factor. If anything, as discussed in Chapter One, disinflation is forcing organizations to take a harder look at costs and their corollaries in product and organizational complexity. It is because markup is no longer a routine solution to high costs that market-driven pricing can add fuel to all manner of cost-improvement activity. When high costs can no longer be absorbed by jacking up the selling price, companies have no other recourse than to work more diligently on cost-improvement activities.

Unfortunately, flawed assumptions around price mask the opportunity and the solution. In many companies, GAAP practices that support obsolete markup pricing remain unchanged.

We are not suggesting that a company drawn to VEP get rid of GAAP. Many aspects of a traditional accounting system can continue to serve the company during a VEP implementation and beyond. The two approaches can live side by side, each making a positive contribution. Activity Based Costing (ABC), discussed briefly at the close of this chapter, is another accounting option.

Instead, we focus on the anomalies in the traditional approach related to VEP in order to understand: a) how GAAP helps to create the problem of exploding variation, and b) what GAAP cannot do to solve it. When that distinction becomes clear, you will know when to set aside traditional practices and when they can remain. Once again, because of its flawed assumptions, GAAP procedures are simply not designed to track levels of organizational complexity or to discover the true cost of product diversification. Let's look.

The Allocation Of True Cost: A New Cost Perspective

Successful companies in today's business ecology recognize the link between product complexity and complexities found in marketing, design, manufacturing, distribution, and support processes. Are there specific costs triggered by this relationship? If so, can they be measured or verified? We already know that traditional accounting, with its cost-masking properties, will not help us. What we need is a new, more practical perspective on cost.

In the 1980s, Japan gave us lessons in building long-term competitive advantage through the systematic identification and reduction of waste—cost. In the 1990s, we began to own those tools and discover new ones. We applied them well, finding and rooting out the high costs of manufacturing—making defects, over-inspection, too much material handling, building inventory. But, as we now realize, there are other costs beyond these, buried deep in the organizational infrastructure—"new" hidden costs. What is new is our awareness of them, not the costs themselves.

The term VEP uses to capture this is "true cost." "True" here is used in the sense of real and actual. True cost is not so much a redefinition of cost as the decision to look for cost in a different place, in its real abode, at its source—the single source of cost, *the part itself*.

VEP defines true cost as "the sum of the chain of costs triggered by the introduction of a new part into the company system." Minimizing true cost is the goal of VEP. In this sense, all organizational activity adheres to a single source of cost: the part itself. This emphasis on true cost is VEP's way of driving our attention to the need for a new cost perspective—one that factors in the full impact of all our business decisions.

The discussion of the "part as the cause of all cost" changes radically when we factor in the impact of additive manufacturing. We, along with you, watch the evolution of this new technology with great interest.

The Origins of Organizational Complexity

In VEP, finding the root cause of variety begins as far upstream as possible, at the initial phases of product conceptualization and design. It is here, as the product moves from concept to prototype and one new part after another is added, that cost begins to accumulate and its various trajectories can be observed ricocheting across the entire organization.

Singular as it may initially appear, it is only when a company discerns waste in this clear and precise manner that the successive layers of complexity triggered by that part can be identified and disassembled. Minimizing these layers is an iterative process. Like a gem cutter chiseling away at the edges of a rough diamond to expose its brilliant core, you have to systematically remove unwarranted variation and needless differences until variety effectiveness is realized. This effort leads us through the dross and debris of antiquated product practices to the smart simple designs that our customers will buy. We begin to make products for profit.

Costs Adhere to Parts: The Part as First Cause

PUI, our case example, began as a one-product company. Within a decade, its product lines began to rapidly expand, with each new product requiring from one to 25 new parts. Now, 70 years later, the company is awash in products and parts; complexity threatens to choke off all profits.

Just One New Part. From the VEP perspective, PUI's unhappy plight got kicked off when the first new part was introduced—and nobody noticed. This is an important point. From the viewpoint of variety effectiveness, the addition of even one new part sets in motion a long causal chain of cost and complication which is almost impossible, if unaddressed, to trace or restrain.

When PUI decided to differentiate its product line, it *inadvertently* triggered a series of negative effects, and a trail of costs followed in its wake—*inadvertently* because its cost-accounting system was not geared to watch for those effects. The first loop of these unintended effects hit the company in four main cost categories—development, materials, production, and facilities. The actual costs spanned a range of activity: from the cost of drawings to the cost of materials management and product defects, from the cost of purchasing new equipment and dies to planning and scheduling, and from utilities cost to the rent or mortgage itself (Figure 3.3).

But PUI's cost list does not end there. Labor hours for designers, drafts people, and other technicians in the Product Development Department increased due to overtime costs and new hires. As parts were fed into the system, drawings and blueprints multiplied. Purchasing began to fight never-ending battles to procure parts and maintain service parts. The shop floor faced a spiraling need for new dies and tooling and a corresponding need for more setups.

Over the years, production processes and methods grew in number and complexity, creating other complications in the flow of production.

As product offerings rose even further, dead stock began accumulating in storage.

Figure 3.3. **Consequences of Adding Just One New Product with Just One New Part**

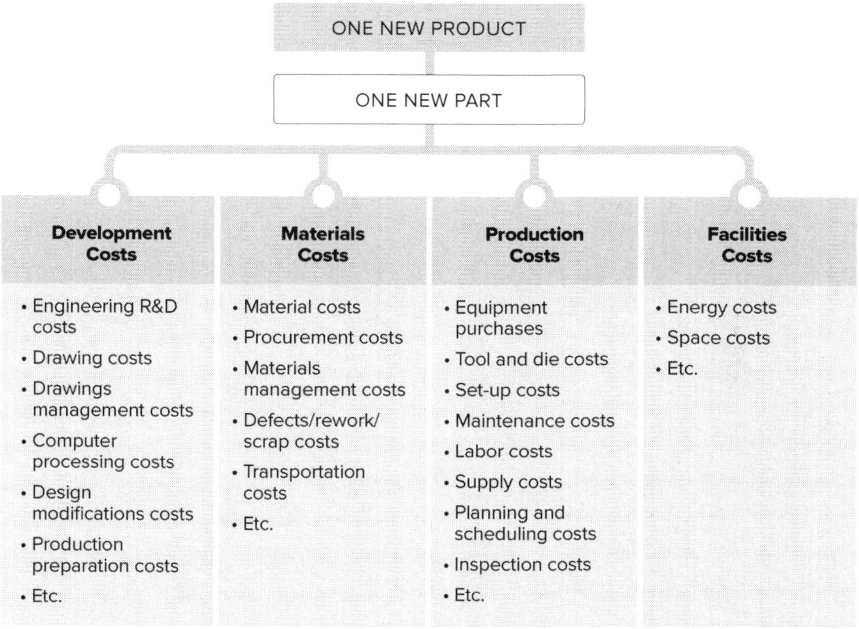

```
                    ONE NEW PRODUCT
                          │
                     ONE NEW PART
```

Development Costs	Materials Costs	Production Costs	Facilities Costs
• Engineering R&D costs • Drawing costs • Drawings management costs • Computer processing costs • Design modifications costs • Production preparation costs • Etc.	• Material costs • Procurement costs • Materials management costs • Defects/rework/scrap costs • Transportation costs • Etc.	• Equipment purchases • Tool and die costs • Set-up costs • Maintenance costs • Labor costs • Supply costs • Planning and scheduling costs • Inspection costs • Etc.	• Energy costs • Space costs • Etc.

In its attempts to deal with mushrooming product lines, the infrastructure at PUI became increasingly cost-burdened. Spiraling complexities made it less likely that their causes could be identified and addressed. Continuing complexity spurred yet further complication and the company became engaged in a desperate struggle to find a box big enough to hold the consequences triggered by adding "just one new part." All this in the face of ever-shortening product life cycles, and sales and profits that never seemed to grow in proportion to the increase in product offerings and parts count variety. The cumulative effect was about to drive an otherwise thriving company out of business.

What happened? Like many companies, PUI had not yet realized that complex products create complex organizations. It had not yet understood that: *All costs adhere to the part.* Once a company grasps this, it must then operationalize this new understanding into practices that help gauge levels of organizational complexity and valuate true costs. VEP provides two mechanisms for doing this—both originally codified in Toshiro Suzue and Akira Kohdate's *Variety Reduction Program*. The first is The *VEP Parts Index* for exposing product and organizational complexity. The second is VEP's *Tri-Cost/True-Cost Model* for differentiating the costs of that complexity.

Measuring Product And Organizational Complexity

We have already seen how the addition of even one new part can trigger dozens of cost categories. In VEP, this simple but powerful understanding is operationalized through The VEP Parts Index, a universal measure of product and organizational complexity.

To illustrate how The VEP Parts Index works, let's observe what happens as a start-up pen company prepares to manufacture its first product series, the J-190s, a ballpoint pen line. The first model, J-191, is a blue-ink pen, simple as products go: eight part numbers at a cost of about $.23 per pen. The part numbers and description are listed below, along with the cost of each part (standard carrying costs already reflected) (Figure 3.4).

Figure 3.4. **Parts List: Model J-191/Blue Pen**

PART #	PART DESCRIPTION	PART COST
1. 1-R-9	One housing	.010
2. J64T	One cap clip-Blue	.035
3. 32-2	One ball bearing	.035
4. 54-BL	Ink-Blue	.023
5. J56-3	One pen tip	.025
6. 5K3	One ink retainer	.070
7. 44-44	One pen guide	.025
8. 77-94	One plug-Blue	.008
	Total unit cost	**.231**

A success, the blue-ink model led to the introduction of a red-ink model the next year: Model J-192. Figure 3.5 shows that the parts list for this second model is identical to that of the first, except for the ink change and corresponding changes in the color-coordinated cap clip and plug. Simple as this product is, however, the ink change resulted in the addition of three new part numbers—or a 38% increase in the company's parts inventory; the five other parts are shared (commonized) between the two models. According to traditional cost-accounting procedures in use by the company at this time, the cost of the two products is identical.

Figure 3.5. **Parts List: Model J-192/Red Pen**

PART #		PART DESCRIPTION	PART COST
1.	1-R-9	One housing	.010
2.	J65T	One cap clip-Red	.035
3.	32-2	One ball bearing	.035
4.	55-RD	Ink-Red	.023
5.	J56-3	One pen tip	.025
6.	5K3	One ink retainer	.070
7.	44-44	One pen guide	.025
8.	77-95	One plug-Red	.008
		Total unit cost	**.231**

The VEP Parts Index: Universal Measure of Complexity

What was the larger effect of adding a red model to the pen company's product line? Again simple as the product is, the addition of the second model led to complicating factors and unexpected costs: the cost of acquiring and handling the three new part numbers, added paperwork needed to order and track parts, separate storage space for each new part, new shelf and bin labeling, the need for additional machine setups when the ink is changed—and the purchase of dedicated equipment if either product gains popularity and demand grows.

    The VEP perspective on these two models targets complexity rather than parts cost alone. This is the purpose of The VEP Parts Index where the parts lists (bill of material/BOM) for each model is laid out in a matrix. It is this matrix format that allows us to appreciate the complexity consequence of our product decision. The matrix displays the number of different part types—as well as the number of times each part number within each of those part types is used. In this case, it shows us that across the blue-ink model and the red-ink model. In other words, we can see which—and how many—parts are shared or communized, and which are dedicated or unique. Once displayed, we apply a simple calculation and arrive at an *indexed figure*. This figure is a relative measure of complexity (Figure 3.6).

Figure 3.6. **VEP Parts Index for Models J-191/Blue and J-192/Red**

PART TYPE	PART NUMBER	MODEL J-191 BLUE	MODEL J-192 RED	TOTAL PART TYPE OCCURRENCES
1. Housing	1-R-9	1	1	1
2. Cap Clip	J64T-BL	1		2
	J65T-RD		1	
3. Ball Bearing	32-2	1	1	1
4. Ink	54-BL	1		2
	55-RD		1	
5. Pen Tip	J56-3	1	1	1
6. Ink Retainer	5K3	1	1	1
7. Pen Guide	44-44	1	1	1
8. Plug	77-94-BL	1		2
	77-95-RD		1	
Total Parts Count		**8**	**8**	**16** **11**

VEP PARTS INDEX (Models J-191 and J-192) = 176 (16 × 11)

The VEP Parts Index is the sum total of all part number counts across their associated BOMs—multiplied by the sum total of the number of times those part numbers as part types occur across models. In the case of our two-pen example, 16 part numbers occur as a part type, 11 times. The multiplied result is our indexed figure—in this case 176. This figure is a relative measure of variety within a given product group. In this way, the VEP Parts Index shows the intersection of part numbers and part types, with the number of occurrences across models under scrutiny. *In other words, variety as a relative measure of complexity is a function of the total quantity of parts handled.*

Within a month, the red-ink model began to outsell the blue-ink, and the company decided to add a black-ink model (J-193) and a green-ink one (J-194). We ask the question again: What was the real impact on the company of adding two more models—the true cost? To answer that question, we look at the four models through the window of The VEP Parts

Index (Figure 3.7) and compare it to the previous one.

Figure 3.7. **VEP Parts Index for Models J-191, 192, 193, 194**

PART TYPE	PART NUMBER	MODEL J-191 BLUE	MODEL J-192 RED	MODEL J-193 BLACK	MODEL J-194 GREEN	TOTAL PART TYPE OCCURRENCES
1. Housing	1-R-9	1	1	1	1	1
2. Cap Clip	J64T	1				4
	J65T		1			
	J66T			1		
	J67T				1	
3. Ball Bearing	32-2	1	1	1	1	1
4. Ink	54-BL	1				4
	55-RD		1			
	56-BK			1		
	57-GN				1	
5. Pen Tip	J56-3	1	1	1	1	1
6. Ink Retainer	5K3	1	1	1	1	1
7. Pen Guide	44-44	1	1	1	1	1
8. Plug	77-94	1				4
	77-95		1			
	77-96			1		
	77-97				1	
Total Parts Count		8	8	8	8	32 / 17

VEP PARTS INDEX (Four J-190 Models) = 544 (32 × 17)

Looking through the window of the new index, we see that the cumulative number of BOM parts has doubled—going from 16 to 32. That makes perfect sense, you say, since the number of products doubled, going from two to four. And the part type occurrences climbed six notches—from 11 to 17. No worry there, you say. But the index itself is behaving strangely. The indexed level was at 176 for two products but jumped to 544 for two more—more than three times the original number and 68% higher than the previous index—not the expected 50%. The situation at the pen company is getting unexpectedly complex.

In the back offices, the search for suppliers of black and green ink has begun: Calls for bids have gone out, new storage areas are being located, and new fields are being entered into the computer. Equipment set-up operators are trying not to think about the messy ink changeovers that will be required. Things are getting just a little bit more congested. But, excited by the increased sales revenue, Sales and Marketing are talking about adding a new model in sparkling pastel colors aimed at teenage girls, and another with a mock Mount Blanc® barrel and cap for the "yuppie wannabe" market.

The VEP Parts Index is not designed to put a judgment or valance on the parts variation it exposes. It simply displays it. The question of whether the resulting variety is positive or negative remains to be answered by the VEP Teams, not by the index. The index is first and foremost a measure and a predictor. As such, The VEP Parts Index is a simple and shrewd yardstick for measuring product complexity and for predicting corresponding complications that *could* follow in the aftermath. In later chapters, you will see how the index is used to support the VEP methodology and how it can be adapted to index other areas of variation triggered by product variety—production processes and control points.

Getting More Complex. Up to this point, our examples have illustrated an application of the index on a highly standardized product with very few parts (many of which are shared)—a ballpoint pen. Observing the expansion of variations across the several models of this simple product helps us appreciate the progressive and accumulating impact of adding new parts.

But the index is also used to expose possible complications associated with complex products. In VEP, a product is considered *complex* when it has some shared parts and many dedicated parts (parts unique to it). If the parts are highly standardized, the likelihood of their being shared across products increases and, as a result, potential levels of organizational complexity drop substantially. But we're getting ahead of ourselves.

Let's look at the more complex product known as the T-144, a pressure control made by PUI, designed for hazardous, explosive applications in oil refineries, chemical plants, and gas pipelines. The VEP Parts Index shown in Figure 3.8 is a partial index, calculated on partial BOMs. It displays 12 part types out of the 25 found across three of the 51 different models in the T-144 series. Other more complex PUI series have models whose BOMs contain from 50 to 200 different parts. The index figure of 1,672 for these partial BOMs begins to suggest the level of complication the company is encumbered by in manufacturing these three models. From a pool of only 12 part types, 76 different parts are used, variously occurring 22 times.

Figure 3.8. **VEP Parts Index: Partial BOM for 3 of 51 PUI Models**

PART TYPE	PART NUMBER	MODEL T-144-06	MODEL T-144-08	MODEL T-144-08	TOTAL PART TYPE OCCURRENCES
Screw	0113-06019	2	2	2	6
	0113-08038	1	1	1	
	0125-10025	3	3		
	0147-08038			4	
	0147-08056	3	3	3	
	0143-40075			10	
Nylon Ball	6219-746	1	1	1	1
Spring Pin	6219-813			2	1
Lockwasher Split No 1/4	2519-25	1	1		1
Explosion-Proof Enclosure	60120-13	1	1	1	1
Insulator	6205-267	1	1	1	2
	6205-387	1	1	1	
Spring	6238-446	1			2
	6238-448		1	1	
Plunger	6240-966	1			2
	6201-298		1	1	
O-Ring	6212-144	1	1	1	3
	6212-281	1	1	1	
	6212-284			1	
Housing	6216-208	1	1	1	1
Mounting Bracket	6222-167	1	1	1	1
I M Label	6233-543	1	1	1	1
Total Parts Count		21	21	34	22 / 76

VEP PARTS INDEX = 1,672 (76 x 22)

One might wonder what it was like in the beginning, as PUI ramped up to launch these exciting new additions. Translated in our imagination, the index is like a TV screen on which we picture product developers scurrying to Drafting with a fistful of drawings, harried process engineers swarming the shop floor in the midst of last-minute ECNs (engineering change notices), procurement specialists upstairs scouring the country for a source for explosion-proof enclosures— located close enough to meet the company's delivery requirements, teams in Cells #930 and #931 scratching their collective heads, trying to figure out where to fit 76 new part numbers in their point-of-use storage. In other words, it's business as usual— complicated, harried, chronically congested, and the unconscious trigger for mistakes, mishaps, and soaring costs.

Not an Exact Measure—But an Exacting One

VEP's Parts Index is not an exact measure. It serves, instead, as an approximation of product cost, with cost as a function of the number of part types and their occurrences by product. It can also be adapted for multiple uses:

- Tracking part types as they are added over time
- Estimating the impact of ECNs on the total parts inventory
- Assessing offerings across diverse markets
- Displaying the intersection of production processes across product models, types, and lines
- Displaying the intersection of control points across control functions and products

Because it is a relative measure of variety, it is possible to use the index to extrapolate the impact of complex products on the organization.

Increasingly, companies realize the traditional measurement approach runs counter to their operational and profit objectives. They know they need a reliable way to quantify—and so validate—the existence of complexity as an organizational problem and gain insight into its whereabouts and causes. The VEP Parts Index reveals the impact of part variation from several perspectives:

- Narrow Perspective: Compute an index on all subassemblies within a model group
- Broader Perspective: Compute an index on all part numbers within a model group
- Widest Perspective: Compute a single cumulative index on the company's entire product universe

In each case, VEP's Parts Index exposes internal part jam-ups and helps track down associated causes.

The Index as an Improvement Driver

One of the challenges of a cross-functional improvement process like VEP is to find a universal measure that the organization can use to gauge its overall progress. Similarly, individual departments need a simple and understandable way to assess their respective success levels. Up until now, no device has been available for estimating, even approximately, the impact of product proliferation on the organization. To attempt to do so with a traditional accounting approach would be so cumbersome and intricate, any benefit such knowledge brings would be nullified. The VEP Parts Index, however, provides such a yardstick. And it becomes more powerful when linked to an improvement-driven approach such as ABC accounting (more about this later in this chapter).

Given that VEP's goal is to reduce parts numbers to a level optimal for the market—and given its central premise that *all costs adhere to the part*—The VEP Parts Index is a useful way to estimate the company's success in minimizing cost as a corollary of parts count. The index is not an exact measure. As we will learn next, however, solid awareness of a significant issue is often more potent than meticulous evidence of an insignificant one—or, as someone once said, "a stitch in time would have confused Einstein."

Now that we've explored The VEP Parts Index and how it can alert us to hidden costs, we turn to the second cost-related mechanism in VEP to understand what form those costs take—VEP's Tri-Cost/True-Cost Model.

VEP's Tri-Cost Model:
The Three Dimensions Of True Cost

We have seen that variety in products and parts can trigger costs and internal complexity. What, then, are the divisions of those costs? What are the costs of complexity? Can this information be codified into a reference model that facilitates inquiry and application? What form does the impact of adding one new product—or a dozen—to the company offerings take? Is there a way for the company to assess what it will spend—or has spent to date—on meeting customer demands? What is the true cost of a single drawing? To find the answers, we look to VEP's Tri-Cost/True-Cost Model (our thanks again to Suzue and Kohdate for conceptualizing this important mechanism).

In a nutshell, The Parts Index makes complexity evident—and The Tri-Cost Model delineates a product's *true* cost. Unlike the traditional labor-burdened GAAP formula discussed earlier in this chapter, this model links directly to the complicating side effects of manufacturing more than one product; it then reconfigures the expenditure of time and assets along the three cost dimensions of function, variety, and control:

- The dimension of function—or what happens as the company fulfills the performance requirements of a single product.
- The dimension of variety—or what happens as the company develops and manufactures diverse products.
- The dimension of control—or what happens as the company orders, tracks, inspects, or otherwise supports the manufacture of a single product or diverse products.

These dimensions are delineated in Figure 3.9.

True cost is the sum of the costs of these three dimensions, expressed as the following formula:

The pen company discussed above, for example, will use this cost equation not to assess the cost of a single product but to delineate the meaning of the index figure (544 in Figure 3.7 above) and its impact on the organization. In conjunction with the Index, VEP's true cost approach looks at the cost consequences of introducing that single new model into an organizational context that is already product-populated. It does not—nor does it attempt to—provide an exact cost reckoning for each product or the dollar amount of those consequences.

We will now examine each of the three VEP cost dimensions.

Dimension 1: Function Costs

A product is a collection of parts. **Function Costs** (F-Costs) are generated as the company furnishes a product with its required functions through parts specifications, values, dimensions, and unit structures. F-Costs are triggered in the design of a single product and also include the materials, processing, methods, and personnel involved in developing, fabricating, assembling, and packing and shipping that product.

Design decisions about product and part values are inextricably linked to the functions those values are contrived to fulfill. Ideally, a product's structural and parts specifications are based, first and foremost, on client and market needs. Frequently, however, this is not the case. Too often, a company fulfills product function through a parts structure that is unduly varied and complex.

Figure 3.9. **Three Dimensions of True Cost Defined**

COST DIMENSION	DETAILS
1. Function Costs (F-Costs)	
...are generated as the company furnishes a product with its required functions through part specifications, values, and unit structures.	*F-Costs* are equivalent to the sum of the labor, material, and processing costs as found in a standard cost-accounting system.
2. Variety Costs (V-Costs)	
...are triggered when a company adds to its stock of design specifications, dimensions, and values in the form of yet another single new part, regardless of whether the addition is customer-initiated or internally caused.	*V-Costs* include added production processes, machines, tooling, fixtures, shelving, space, etc.—all that is required to fulfill the demands of multiple products, product types and models—and encompasses the cumulative consequence of diversity on the entire system.
3. Control Costs (C-Costs)	
...refer to tasks and information transactions geared to control, track, manage or otherwise support a product and its parts as they move through the system to the end user.	*C-Costs* refer to the myriad of transactions the company undertakes to design, acquire, order, receive, inspect, track, store, retrieve, count, handle, maintain, or otherwise support and manage the manufacture and sale of its offerings.

Many companies facing high Function Costs have attempted to redress this condition through value engineering/value analysis (VE/VA). The intent of VE/VA is to minimize the cost of realizing product function. This can be very effective in minimizing Function Costs on a case-by-case basis—but may cause a simultaneous rise in Variety Costs (more about this in Chapter 4).

As a category, F-Costs may be taken as the sum of the labor, material, and processing costs as found in a standard cost-accounting system. Seventy years ago, these costs accounted for more than 90% of the cost of a product. Today, in many businesses, they range as low as 40%.

Dimension 2: Variety Costs

Variety Costs (V-Costs) exist in companies that have a diversified product line. These costs are triggered when a product line expands, even if only one attribute of a single part is altered, however slightly. Regardless of whether such changes are customer-initiated or internally triggered, the result is a variety cost.

Variety Costs, then, are the costs that arise from product diversification. These costs include the additional production processes, machines, tooling, fixtures, etc., required for the enterprise to fulfill the needs of additional products and multiple product types and models. This variety factor applies across product lines and encompasses the cumulative consequence of diversity on the entire system. The engine of complexity is variety costs. Overall, they are estimated to constitute 25% of the total cost of the product.

An organization may adopt several strategies for dealing with the demands of a proliferating product line. For example, the company may pursue standardization as a remedy. In this event, engineers seek to avoid product designs that call for dedicated or specialized parts and strive, instead, to design parts that can be shared across as wide a range of products as possible—a critically important tactic as far as VEP is concerned.

In another case, the company may elect to meet the need of processing the differing parts by investing in additional equipment and/or in flexible manufacturing equipment (e.g., Computer Numerical Control [CNC] machines). Sooner or later, the limits of this equipment are reached; the company may then decide to purchase machines and tooling that promise even greater range and flexibility. In such a case, it is not unusual for the company to pass on the cost of this decision to the customer in the form of higher prices.

Along the same lines, a company seeking to meet the needs of ever-burgeoning product lines through equipment purchases can find itself in a double bind. If it continues to allocate resources for new tooling, it may be forced to forego other investment opportunities. Alternatively, in light of a mushrooming parts count, the company could discount automation as a long-term solution and decide not to invest in it—and inadvertently forfeit money-saving benefits. In both cases, exploding variety renders these options expensive, and neither of them provides a durable solution to the problem.

Dimension 3: Control Costs

Control Costs (C-Costs) refer to the indirect tasks and information transactions that support the other two cost categories—Variety Costs and Function Costs. C-Costs include costs for design, drafting, ordering, buying, inspecting, transporting, storing, and maintenance. In scope and variety, they are roughly equivalent to *overhead*.

Each transaction and *each* informational exchange is a control point. The wages of all non-operational personnel involved in these activities are also C-Costs. For example, what are the cost corollaries of a single drawing? Let's track the cascade of control points triggered by that.

Once the design is generated by the designer, it winds its way through the company to drafting, then prototype, pilot, and revision. There are many other aftereffects of a single drawing, viewed from the opposite end of the plant:

Group 1: Raw material/parts are sourced, ordered, and paid for.

Group 2: Raw material/parts are received on the dock, inspected, handled, and stored.

Group 3: Raw material/parts are retrieved and checked for quality after processing and/or assembly.

Group 4: Finished goods are packed and shipped to the customer.

Other activities associated with this cost category include: production scheduling, data processing, and inventory control—which attempts to answer the question: How much should be kept on hand?

Control Costs, then, delineate the overhead expense incurred to support products, parts, and processes. These are the costs incurred by people—the people in Design, Operations, Material Handling, and other support functions. Control Costs also refer directly to the myriad of transactions that people in the company undertake to acquire, receive, inspect, track, store, retrieve, market, sell, and support products.

Literally dozens—often thousands—of control points adhere to a single product, or even a single part. Control points are single product- or part-specific actions and transactions that touch every department. Taken in their sum, they create a workplace that is hyperactive with detail and swimming in minutiae. The population size of control points found in an organization goes a long way in making and keeping it clogged with complexity. However committed a company may be to world-class excellence, its progress will eventually get stymied if control points are not themselves brought under control and minimized.

The VEP Cost Pie

Seventy years ago, F-Costs accounted for 90% of total manufacturing cost, with V- and C-Costs sharing the remaining 10%.

But that was then and now is—different. World War II brought huge changes to the meaning of success in business, with product diversification increasingly viewed as a competitive strategy. With those changes, V-Costs and C-Costs began to take ever-widening shares of the total cost pie. By cost category, the percentage share of total cost generally assignable to each is as follows (Figure 3.10):

F-Costs: estimated at 40% of the total cost of the product

V-Costs: estimated at 25% of the total cost of the product

C-Costs: estimated at 35% of the total cost of the product

Figure 3.10 Divvying Shares: F-Costs, V-Costs C-Costs

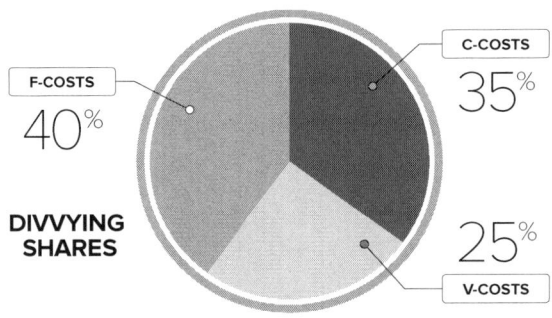

The shares of the costs shown above are approximations at best that can fluctuate widely according to the type of industry and markets. In specialized industries like brick manufacturing, for example, F-Costs generally hover around 15% because the product is highly standardized; V-Costs, on the other hand, may be elevated, possibly as high as 55%, due to the costly chemicals required to meet the range of brick values needed in that market.

In highly-standardized sectors such as automotive, mass-production electronics, and other volume industries, F-Costs reach as high as 50% to 70%. V-Costs, by contrast, account for as little as 20% of total cost, with

C-Costs as low as 10%. On the other hand, special-order industries like furniture, heavy earth-moving equipment, and customized electronics trigger C-Costs that climb to 45%.

One further word on the Control Cost share. These generally stay steady in the 30% to 40% range due to the high degree of similarity across industries in standard bureaucratic and administrative procedures. Exceptions include businesses dealing with highly-regulated government contracts, which require significantly more documentation and inspection activity than does private-sector work. In these instances, if *all* control points were carefully tracked in terms of VEP's definitions, C-Costs might soar into the 50% level or higher.

Control Points and ABC Accounting

It is the variety in products, parts, and processes that results in excessive control points. We must also point out that many control points are triggered by the outdated accounting practices described earlier in this chapter. This labor-burdened approach often overlooks or ignores wasteful variety; it may even encourage or create it because:

- It spends vast quantities of time and energy measuring the wrong things in the wrong way.
- It neglects to measure the right things efficiently and effectively.
- It directs the work force, as a consequence of the above, to pursue irrelevant, and often detrimental, objectives.

Part of the beauty of VEP's Parts Index and Tri-Cost Model is that they do not encourage companies to get caught up in ascertaining costs in terms of exact dollar amounts, one of the problems with the traditional cost-accounting practices of GAAP. So much time and effort are spent in getting to-the-penny-exact figures that the purpose of accounting, to begin with, is lost. In our view, a more relevant approach is activity-based costing (ABC). While it is not our purpose to promote or explain the ABC approach, we will mention two ABC factors that align nicely with VEP objectives:

- ABC does not make a strict differentiation among material, labor, and overhead costs as defined in traditional accounting principles.
- ABC groups them all as indirect costs and then seeks to link each of them with a specific source activity (known as a cost driver).

With ABC, companies select the cost drivers most relevant to their process and their improvement objectives and begin to track those, for example: the number of purchase orders, customer orders, engineering change notices, material moves, machine setups, tools issued to the shop floor, product insertions, manual soldering tasks, products shipped, and so on. A company using an ABC system learns to reassign costs to specific products, in correct proportions—costs previously allocated as indirect. In ABC costing, for example, indirect costs are separated into procurement, production, and support activity.

Awareness Is Everything

The goal of VEP is to maximize customer selection by reducing the negative effects of product variety, strengthening the positive effects, and preventing the return of negative variety. That means optimizing external variety (varying products so that customers buy them) and minimizing internal variety (how the organization provides those varying products).

In this chapter, we introduced two of the tools central to this effort. First, The VEP Parts Index helps you determine whether your company shelters negative variety. Second, VEP's Tri-Cost Model, helps you ascertain how negative variety expresses itself in your organization. Between the two, you can identify and validate the need for action. While you may never succeed in eradicating all traces of negative variety, you can expect to find a balance point between it and positive variety. That balance point is called effective variety. Achieving this balance does not necessarily mean that the enterprise makes more revenue—or even saves more money. It means: The company *makes more profit.*

As much as a company needs reliable accounting procedures, it is equally urgent to cultivate a dynamic awareness in the organization that negative variety creates cost. The VEP Index and Tri-Cost Model are designed to promote that awareness and trigger further insights so that you and your team recognize the existence of negative variety in your company. Once you recognize it, you will move to action—improvement action. You will do something about it.

Next we will look at the triggers of cost and complexity, caused by new product expansion—the triggers of the variety explosion.

CHAPTER 4
Negative Variety And Its Policy Triggers

There's no getting around it: When you ship your product, you ship your corporate decisions—your policies and practices.

The introduction of new products and services is at the top of the competitive agenda of practically every company today. This agenda is driven by three forces. First, markets have been internationalized—competition has become global. Companies can no longer count on growth and security from success on a regional or domestic scale alone. Even if companies thrive in domestic markets, the threat of an invasion by foreign competitors is never far from view; protective steps must be taken. To keep foreign competition at bay, Cummins Engine at one point spent over a billion dollars (three times the market value of its stock at the time) to expand its engine products simultaneously on three fronts, followed within four years by slashing prices on its new products by as much as 30%—all to fend off imports.

    A second powerful stimulus for new products lies in technology breakthroughs. Science continually pushes the edges of the innovation envelope in nearly every industry. Advances in technology now make it possible for a company to give consumers the highly differentiated products they increasingly demand. This is nowhere more vividly seen than in the computer industry where advances in technology have resulted in the introduction of a flood of new-to-the-world products. In addition to innovations to your basic computers, monitors, and printers, the market offers an ever-widening range of laptops, tablets, GPS units, iPads, iPhones, netbooks, smartphones, e-readers, plus legions of peripherals and accessories.

By the same token, scientific and engineering breakthroughs can spread like wildfire through the R&D departments of the world, turning yesterday's product triumph into tomorrow's has-been. Product cycles were once measured in decades. But today's personal computers and smartphones become outmoded in just months. Every victory is accented by the lurking awareness that your competitors just may be able to purchase or license the same technology, discover it themselves or develop its next generation through their own R&D. In short, the same technology that gives you market dominance today may turn around and bite you tomorrow.

The third force driving the new product agenda is the consumer. Consumers—no longer passive bystanders to the process—have become knowledgeable and demanding. Markets are highly segmented *and* the buying public expects more and more for its purchasing dollar. Customers want choices on much deeper levels than ever before. Where price and basic performance used to be sufficient reason to buy, consumers now look for products with the "right life-style values." Gillette, manufacturer of a wildly successful razor line, never gouges its consumers on price. Instead, it makes demonstrably superior products and innovates constantly—too rapidly for a competitor to copy. "We don't sell products. We capture customers." The very act of consuming sensitizes customers to the possibility of more product differences.

These global forces shape and amplify the explosion of product variety.

Variety Explosion: Unintended Consequences

There is another giant hiding in the corporate closet—one that seems coterminous to product proliferation but in fact has a life and an impact of its own. The riotous growth in products is accompanied by a simultaneous explosion in parts inventories. This giant has been in hiding for a very long time. The result? A level of internal complexity and congestion unparalleled in the history of business and industry.

To a great extent, this condition is both the beneficiary and victim of accelerating capabilities in computer technology. The computer—and the canny software that supports it—allows companies to generate huge annual crops of new products with levels of speed, ease, and ingenuity previously unthinkable, limited only by human imagination. The same computers provide organizations with a capability to store and manipulate immense amounts of information related to the design, marketing, and production of these products. In the act of doing so, vast new quantities of data are added to an already-swollen base.

But the computer was not designed to manage itself or its output. Few companies are prepared to cope with the sea of data or the glut of diversified products that result. Even fewer factor in the secret costs of introducing all those exciting new products. Nor are they prepared to absorb these costs, costs that continue to drain even companies with sterling, super-lean, super-visual production systems. The cruelest blow is the dawning realization that the customer does not always want the high level of choice these businesses work so hard and spend so much to provide.

Nissan, a top Japanese automaker, is a good example. Let's look at some of its past indiscretions. In the late 1990s, when frenzy over new computer capabilities was peaking, Nissan's quest was to build cars in ever more sizes, colors and features to satisfy the presumed whims of the world's drivers. Its car model variations ballooned to more than 2,200—with 437 different kinds of dashboard meters, 1,200 types of floor carpets, and 300 different kinds of ashtrays. In one model alone, Nissan offered 87 types of steering wheels. Under the hood, customers could find (if they cared to look) one of 110 types of radiators, all held together with over 6,000 different fasteners. The number of possible combinations for unique products bordered on astronomical. When Nissan took a closer look, it discovered this: a full 50% of its 2,200 model variations contributed only 5% to total sales.

During this same period, swelling costs tied to product variety were not confined to Nissan—but found in all eleven of Japan's automakers. Similarly, Matsushita, the Japanese electronics giant, was saddled with 6,000 models of stereos and portable tape players. The variety explosion at these OEMs (original equipment manufacturers) rippled through a chain of suppliers that staggered to respond with a supporting medley of components. Calsonic, for example, one of Nissan's biggest suppliers, was forced to make no less than 2,000 kinds of mufflers alone—the majority of which were used at a rate of a less than three units per year.

In the fat years, companies could afford such excess. Now those same companies learned that such an array of choices is certainly too expensive and may even be ultimately fruitless.

Negative vs. Positive Variety

Unrestrained variety in designs and models is not a requirement for increasing market share. In fact, much of what passes for increased customer selection is, in reality, needless variation. Nissan's 87 steering wheels is a good example; it is the rare customer who buys a car because of the steering wheel. When a choice is pointless, it is negative.

Negative variety adds cost, not value. It burdens the corporation with complexity and expense. Ultimately, the customer pays.

Positive variety, on the other hand, adds value, increases sales, and can cut costs.

Positive variety is customer-driven; it increases market share because it gives customers what they want—really want. Because positive variety is driven or pulled by the customer, it fuels growth in sales revenues. It makes, sustains, and grows markets. In this light, continued product proliferation becomes a corporate imperative. We want to make this perfectly clear: The era of the universal "any-color-the-customer-wants-as-long-as-it's-black" product is over. Today, businesses cannot survive—much less enjoy consistent profits—without continually bringing out new products.

But do car buyers really need (or want) 437 different dashboard meters from which to choose? Does such an extended range of choice ever actually clinch the sale of a car? Will the fact that 6,000 different fasteners are used across a product line ever help a consumer make the choice to purchase a car from one company over another? It is hard to imagine even one instance where this could be true.

Even when variations are identified as customer-driven, they must genuinely be so. Authentic customer-driven variation occurs as the result of careful market research, using proven market techniques. In its absence, variation can sneak in as someone's notion of a "good idea." The annals of sales and marketing are littered with such costly notions. Borden Foods brought new meaning to the term *niche market* with its 3,200 different varieties of snack foods. What organization, we ask, could possibly manage this truckload of differences? Borden folded in 2005.

Joe Pine, author of the book, *Mass Customization,* was exact when he said, "Customers don't want choice. They just want exactly want they want. Profitable product-line expansion requires that you not overload your customers with too much information and too many choices, forcing them to spend too much time figuring out if you have what they need."

Positive variety only, please! Everything else—strategy, design, operations, and all other levels of technology and systems—flows from that. The way to ensure maximum profit is to build the principles of variety effectiveness into all layers of your organization.

The Balance Point: Effective Variety

Effective variety is not the static endpoint of a journey. Instead, it is the act of finding the ever-shifting point of balance that is right for your organization—between what your customers want and the internal requirements of running a successful business.

Understanding the distinction between positive and negative variety—and their respective causes—is critical to striking that balance point. The difference is this: Positive variety is externally driven by the *customer*—directly linked to verifiable customer interest or demand. Negative variety is everything else; it is the result of the *way the business operates*—its policies, practices, and systems. In other words, negative variety is *internally-triggered*. Effective variety is the balance point between the two. Defining that distinction and making it active in your enterprise is the purpose of this book.

Just remember: Introducing customer-driven products (positive variety) is only half the formula for achieving effective variety. The other half is minimizing those aspects of product variety that are not linked to a known, valid, or anticipated customer need, expectation or want. The absence of these links is caused, without exception, by the way the organization operates, however inadvertent or accidental. We call these factors as a group: *internally-triggered variety*, which is also always negative. All variety—customer-driven or internally-triggered—generates cost and complexity. But internally-triggered variety exacts an unacceptable trade-off between cost and return.

Though this point of balance is never static, the goal remains the same: Root out the causes of negative variety and find the least-sum means. These causes are unavoidably linked to specific company practices. Virtually every part of the organization—Marketing, Sales, Design Engineering, Process Engineering, Operations, Purchasing, IT Systems, and others—can make either a positive or negative impact on effective variety through the formal and informal policies it follows. It is these policies that help or hinder efforts to achieve effective variety; they are the focus of the rest of this chapter.

The Policy Roots Of Negative Variety

Negative variety is defined as: *Any variation that is not in response to a validated customer need or request*. Unintended as it may be, negative variety is always triggered by the way a company does business; it is all internally-triggered. Internally-triggered variety originates in—and further burdens—

the inner workings of a company, making them more complicated, extending lead time, lowering margins, inflating costs, and, all too often, swelling prices. Everybody loses.

By definition, internally-triggered variety (negative variety) never adds value for the customer and always adds cost. *But it is also not always avoidable.* Achieving effective variety is a two-fold task: minimizing negative variety and enhancing positive variety—the kind that is genuinely customer-driven (Figure 4.1).

Figure 4.1. **Effective Variety: An Ever-Shifting Balance Point**

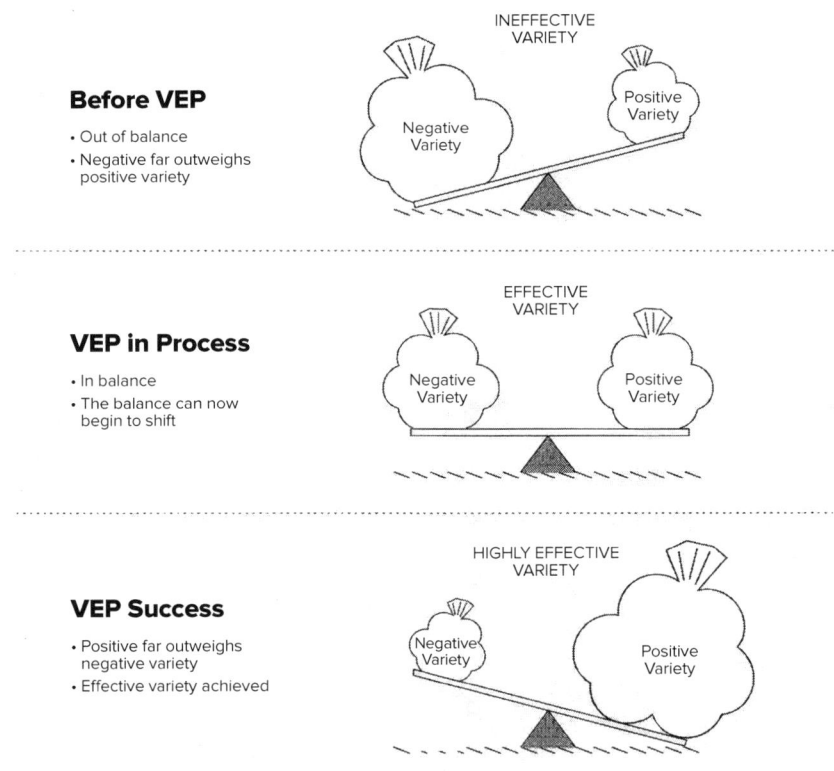

The challenge is to find *and* maintain a balance point between maximal value and least cost. In the process, the organization is de-complicated. Here we include not just proliferating parts but the after-effects we have repeatedly mentioned—exploding production and nonproduction processes, dies, fixtures, control points as well as complexities in distribution, relationships with clients and customers, and in marketing and sales transactions.

This need to de-complicate the enterprise extends to all departments and all functions and refers not only to simplifying products, parts, and processes—but also to refining and improving specific behaviors and practices which encourage or provoke needless variety and, hence, cost.

We group all such behaviors and practices under a single company term: *policy*. Whether officially sanctioned or not, policy refers to any rule, logic or principle that has become standard or usual within the organization. Formal or informal, policy reflects what is expected and, to a large extent, what is done. In the remainder of this chapter, we examine the policies across the following five functional areas and discover seventeen triggers of negative variety (Figure 4.2):

- Accounting
- Marketing & Sales
- Product Development
- IT/Data Systems
- Operations

Figure 4.2. **Negative Variety: 17 Policy Triggers**

ACCOUNTING	MARKETING AND SALES	PRODUCT DESIGN	IT/DATA SYSTEMS	OPERATIONS
1. Supplier Base Selection	4. Response to Customer Requests	8. Products Diversification Approach	15. Product Documentation, Classification Systems, and Computer Support	16. Lot Sizing
2. Make vs. Buy Decisions	5. Margins, Product Pricing, and Discounting	9. Products as Separate Entities		17. Capital Equipment and Process-Improvement Justification
3. Overhead Cost Allocation	6. New Product Market Requirements	10. Long Lapses Between Products		
	7. Cost Targets/ Cost Reductions	11. Different Designers/ Different Design Concepts		
		12. Cost Targeting in New Product Development		
		13. Technical Solutions		
		14. Product Life-Cycle Decisions		

We will look at them one by one, identifying and analyzing policies in each that hinder positive variety. At the end of each analysis, we suggest improved policies, geared to promote the VEP goal of effective variety. These improved policies are presented in grid form at the end of each section, unless otherwise noted.

A. Negative Triggers in Accounting Policies

Accounting practices can be dangerous variety proliferators. Here are three where VEP thinking is crucial.

1. Supplier Base Selection
2. Make vs. Buy Decisions
3. Overhead Cost Allocation

1. Supplier Base Selection. In any given year, a company may have hundreds of active suppliers—multiple suppliers for each purchased part in order to keep the piece price down and to protect against line shutdown in case of supplier calamity (e.g., a fire at one supplier's factory). Over time, however, this competitive practice can create adversarial relationships between company and suppliers for several reasons. First, competing suppliers often catch wind of deals that "pet" suppliers strike with favorite buyers. Second, calamities rarely, if ever, occur. In any case, it is hard to imagine a disaster serious enough to justify the dozens of different suppliers some companies retain for a single commodity. Excess suppliers equal excess cost.

We can see the burden of this policy, for example, in Accounts Payable, where supplier payments are fulfilled by a burgeoning number of employees. An improved policy would promote supplier-customer partnerships, requiring that "we put all our eggs in one basket and then watch that basket." Adherence to this guideline reduces sourcing activities, ranging from telephone calls and faxes, to delivery and quality standards, drawings, and checks in payment.

A word of caution. If a company should choose to adopt this policy, the supplier reduction target must be aggressive enough to force a fundamental revision in how the company deals with suppliers. While impressive in some respects, for example, a 20% reduction in your supplier base may not result in a substantive change in your sourcing strategies.

Since maintaining a dynamic equilibrium between parts and processes is a long-term goal of VEP, another improved policy would require that a new supplier be added only if an existing supplier providing the same item is first removed.

Improvement recommendations will be in a shaded area at the end of each policy segment, unless noted otherwise.

Figure 4.3. **Improved Policies for Trigger 1: Supplier Base Selection**

> **Trigger 1**
> **IMPROVEMENT RECOMMENDATIONS: SUPPLIER BASE SELECTION**
>
> **Identify flawed supplier policies that make supplier selections based on:** a) cost-per-piece only; b) purchasing pets; and c) calamity protection.
> **Replace them with the following improved policies.**
>
> **Improved Policy 1.1**
> Establish partnerships with best suppliers and provide 90% of the business for a particular commodity to the primary supplier that can best serve the company's requirements for service, quality, price, and delivery.
>
> **Improved Policy 1.2**
> Set a target for supplier reduction that is aggressive enough to require a genuine change in policy (for example, reduce supplier base by 40% in the next year).
>
> **Improved Policy 1.3**
> Do not add a new supplier to the supplier base without directly offsetting another supplier that provides the same commodity.

2. Make vs. Buy Decisions (Cost Per Part). Far too often, decisions regarding the *manufacture of parts vs. purchase of parts* (make vs. buy) are based on the cost per part—as determined by internal costing methods that apply overhead based on direct labor costs. As a result of this practice, subcontracting and outside purchasing can become a company's first choice for handling as many production operations as possible.

In other instances, the make/buy choice comes down to whether a process already exists in the plant. If it doesn't, common practice dictates: Let someone else make it. On paper the company may appear to be saving money; but people and machines often stand idle and could be used to make those same parts.

The same cost issue (cost-per-piece) can cause a company to purchase parts in a variety of materials instead of standardizing on a single, superior material that meets the needs of multiple applications. For example, a plunger or screw may be stocked in several materials— brass, brass/nickel, aluminum, and stainless steel. In the interest of value engineering, each part is matched to the nominal environmental specifications for its function.

This often creates dis-economies of scale (three years of stock-on-hand) for each material variety.

By choosing a more robust material (the one with the broadest range of applicability), the quality of all products will be improved while true costs are reduced (including such control costs as drawings, purchase orders, expediting, and billing). This is part of a larger set of considerations found in the VEP methodology and discussed in detail in Chapters Eight and Nine.

Figure 4.4. **Improved Policies for Trigger 2: Make vs. Buy Decisions**

Trigger 2
IMPROVEMENT RECOMMENDATIONS: MAKE VS. BUY DECISIONS
Identify flawed procurement policies that result in automatic parts out-sourcing due to: a) cost-per-piece accounting; b) restricted internal capacity; c) limited process capability; or d) "just not feeling like making it in-house." **Replace them with the following improved policies.**
Improved Policy 2.1 Design parts to favor existing or planned in-house part production process capability.
Improved Policy 2.2 Where a choice of raw materials exists, design parts to standardize on the most robust material that satisfies function across many products.
Improved Policy 2.3 Let raw material cost, capacity, and process capability be the only constraints in determining make vs. buy decisions. Do not consider L+M+O overhead costing.

3. Overhead Cost Allocation. As discussed in Chapter Three, the traditional approach to allocating overhead costs still used in many companies assumes that a product with heavy direct labor should absorb a lot of overhead. But that product might also have few parts, making its *true* total cost lower than that of another product that requires less labor but contains many parts—some of which may sit in stock and rarely turn over. This was the point in Chapter Three of comparing Product 8, with 46 parts, and Product 11, with only 29 parts with a heavier overhead burden because of its higher labor content (Product 8 at 10 minutes vs. Product 11 at 21 minutes; see Figure 3.1 in Chapter Three).

Many companies do not realize the relationship between parts count, stock-on-hand, operating expense *(control costs)*—and the fact that operating expense, in great measure, reflects the total of parts, warehoused and shipped, per year (as opposed to total labor per year). The widespread use of traditional allocation systems shows this. Operating expense in an assembly factory, for example, is very much a function of parts variety (as well as of the number of processes). Therefore, an allocation system that references parts count, rather than direct labor, would provide a far more reliable gauge of costs.

Often, an allocation system that references parts count also recognizes turns in parts inventory as a macro cost measure; in many such best-practice companies, turns in inventory can reach rates as high as 30-50 times a year. On the other hand, inventory turns in companies struggling to address the issues discussed in this book range from three to four a year.

Such companies fail to recognize that overhead associated with inventory is actually time-based: The longer we perform activities associated with parts, the higher the control costs associated with them. A secondary inventory metric to consider is the quantity of stock-on-hand as it deviates from the previous fiscal quarter.

Figure 4.5. **Improved Policies for Trigger 3: Overhead Cost Allocation**

Trigger 3
IMPROVEMENT RECOMMENDATIONS: OVERHEAD COST ALLOCATION
Identify flawed accounting policies that: Allocate product and inventory costs based on an obsolete Labor + Material + Overhead (L+M+O) formula. **Replace them with the following improved policies.**
Improved Policy 3.1 Develop a costing system that allocates overhead based on: a) number of parts in a product, and b) inventory turns per part.
Improved Policy 3.2 Do not use a direct labor-based overhead cost structure for determining inventory costing, except in valuating it for taxes.

B. Negative Triggers in Marketing and Sales Policies

Marketing and sales functions also harbor hidden triggers of negative variety. Here are four worth your consideration, with VEP in mind. (Please note: The triggers are numbered in sequence across the five organizational areas.)

  4. Response to Customer Requests
  5. Margins, Product Pricing, and Discounting
  6. New Product Market Requirements
  7. Cost Targets/Cost Reductions

4. Response to Customer Requests. Some organizations have a reputation, both externally and internally, for "doing anything for the customer—no matter what."

As one manufacturer relates, "We build quantities of 10,000 once. And we build quantities of one, once. You name it and we do it and somehow get the order out the door. But we usually get stuck with lots of new parts, equipment, and paperwork. Last week, we built a product we hadn't seen in two years. It took us three days to unearth the special fixtures used to calibrate it. We never did locate the right drawing, so it took a manufacturing engineer two days to mark up a standard product drawing for us to meet the special specs. This happens all the time. It seems as if every engineer wants to build the product just a little differently, and every new project that hits the floor needs just one or two different jigs or fixtures."

Figure 4.6. Improved Policies for Trigger 4: Response to Customer Requests

Trigger 4
IMPROVEMENT RECOMMENDATIONS: RESPONSE TO CUSTOMER REQUESTS
Identify flawed customer-response policies that: a) give customers whatever they ask for; b) provide indefinite product support; c) obsolete a product only when internal complaints get loud enough; or d) obsolete a product, arbitrarily, in response to periods of product non-activity. **Replace them with these improved policies.**
Improved Policy 4.1 Determine product obsolescence, based on the tracking of product life cycle.
Improved Policy 4.2 Set up a continuous product profitability feedback loop that defines and identifies trigger points for retiring unprofitable offerings.
Improved Policy 4.3 Determine and publish an obsolescence timetable (for example, three-year notice before obsolescence; five years for maintenance and repair).
Improved Policy 4.4 Provide quotes for customer-specific products that state minimum annual usage amounts required for continued company product support.
Improved Policy 4.5 Provide all sales and marketing personnel with on-going education on effective vs. ineffective variety.
Improved Policy 4.6 Implement these improved policies as an on-going and required parts of the marketing function.

Companies that provide open-ended support for customer-specific products often experience serious disruptions in service to their other customers. Typically, such support is not linked to contracts that state pre-agreed customer usage levels—nor do these documents contain language that states the company's policy on product obsolescence. In reality, far too many companies do not have such a policy and continually pay for that oversight.

In addition, salespeople inadvertently add to a company's cost woes—if they have not been educated about the link between giving the customer everything and increased parts inventory levels. This comes as no surprise since traditional accounting practices provide no motivation for salespeople (or engineers) to consider parts count and excess variety as cost factors in the customer/order interface.

5. Margins, Product Pricing, and Discounting. Understanding the *true cost* of your products (as defined in Chapter Three) is indispensable to setting the right selling price. Without that, you might assign a much lower selling price to a high true-cost product—versus a product that is a breeze to plan, manufacture, and support. If you have not identified a product's true cost, you are pricing blindly.

One marketing vice president, complaining about her company's current pricing policies, put it like this: "In management meetings, I am constantly called on the carpet to get profit margins up. The only data my people have to go on are the cost figures in the computer. So we use those to determine price. But we don't know what things actually cost because, for starters, we don't know which parts are turning over and which aren't."

Product discounts are often made in a similar information vacuum. A company may decide, for example, that a price reduction (discount) on its more mature products will spur new sales and gain profit. In reality, additional revenues may not offset the costs of continuing to support those products. In keeping with VEP, it may be wiser to retire them if they cannot be made more variety effective.

Another common (and often unstated) policy is to price products according to market value: What the consumer will bear. If the company sets prices above market value, however, market share can be lost. Other troublesome results ensue if factors related to L+M+O (labor + material + overhead) are used to drive the company's pricing and discounting decisions. For example, because traditional costing can generate wide swings of over- and under-valuing per piece costs, pricing policies based on them are equally unreliable.

In addition, minimum order releases for customers—a vestige of the traditional economic order quantity (EOQ) policy—dictate that a minimum number of items must be ordered so that the company can justify the costs of filling an order (e.g., cost of machine changeovers). This practice leads customers to buy more than they need and the company to respond to false demand while other real demand has to wait for production resources.

Figure 4.7. **Improved Policies for Trigger 5: Margins, Product Pricing, and Discounting**

Trigger 5
IMPROVEMENT RECOMMENDATIONS: MARGINS, PRODUCT PRICING, AND DISCOUNTING
Identify flawed pricing policies for setting price based on: a) profit margin as calculated by the traditional L+M+O formula; b) what the market will bear; and c) economic order quantity. **Replace them with the following improved policies.**
Improved Policy 5.1 Base profitability on the comparison of net selling price to total cost, in which material and overhead are allocated on a part/product basis.
Improved Policy 5.2 Hold prices of mature products constant in mature markets.
Improved Policy 5.3 Figure product price based on market forces—not L+M+O.
Improved Policy 5.4 Provide product discounts based on total order quantity, without regard for minimum releases (EOQ).

6. New Product Market Requirements. Many companies do not have a well-defined strategy for evaluating new product requests. Instead, Marketing or Sales initiates a new product request and supports it with some data on market size, projected sales, competition, and a laundry list of product specifications and cost constraints. In the early stage of new product review, consideration is rarely given to separating the functions desired by the market from how they can be achieved. Questions of product structure, production processing, and support methods are seldom raised.

In some companies where evaluation of consumer needs is inadequate, products are often developed piecemeal in passive response to individual demand along a random timeline. Without a unifying approach, new products often cannibalize the sales of existing ones. As a consequence, economies of scale are reduced and sales volumes split between old and new products. In almost all such instances, parts and processes proliferate and costs increase for the older lines. Negative variety explodes.

Cannibalization can hide deeper problems. Instead of filling defined product voids within existing markets, a company may develop new products in order to achieve one of these two short-sighted outcomes:

- Another "me-too" product for markets already well served by its own existing products.
- A product with a higher profit margin in an existing but competitive target market.

Figure 4.8. Improved Policies for Trigger 6: New Product Market Requirements

Trigger 6
IMPROVEMENT RECOMMENDATIONS: NEW PRODUCT MARKET REQUIREMENTS
Identify flawed new product policies that result in the introduction of: a) products into new markets without sufficient research; and b) so-called "new" products that cannibalize existing products or are simply revisions of them. **Replace them with the following improved policies.**
Improved Policy 6.1 Develop a well-defined new product development strategy that includes screening criteria for evaluating new product requests.
Improved Policy 6.2 Develop new products that fill voids within previously targeted markets instead of products that are redundant.
Improved Policy 6.3 Sell existing products to any market but do not undertake new product development for markets not specifically targeted in your strategic plan.

7. Cost Targeting/Cost Reductions. Many companies develop new products for existing markets which do not constitute a new technology or a first-to-market strategy. In these cases, product cost is the driving force in an effort to achieve an attractive market price and undercut the competition.

This policy often causes designers to favor lower-priced new parts over existing parts that may be slightly more expensive. Other costs associated with the requirements of new parts—more processes, more support tasks, numbers of parts, and expected inventory turns—are rarely considered.

Figure 4.9. Improved Policies for Trigger 7: Cost Targeting and Cost Reductions

Trigger 7 IMPROVEMENT RECOMMENDATIONS: COST TARGETING AND COST REDUCTIONS
Identify flawed cost-targeting policies that: Trigger new products based solely on a traditional lowest-cost profile and not on a true-cost profile. **Replace them with the following improved policies.**
Improved Policy 7.1 Base cost targeting on two factors: 1) number of parts in the product; and 2) number of inventory turns of parts in the product (with an annual turn of four equaling unity).
Improved Policy 7.2 Avoid entering markets solely on the basis of a lowest-cost product.

C. Negative Triggers in Product Development Policies

Product development policies—or the lack of them—are fertile ground for negative variety. For the most part, these are related to one central problem: the lack of a unifying product development approach. Our discussion groups them in seven categories:

8. Product Diversification Approach
9. Products as Separate Entities
10. Long Lapses between Products
11. Different Designers/Different Design Concepts
12. Cost Targeting in New Product Development
13. Technical Solutions
14. Product Life-Cycle Decisions

Recommendations on improved practices for these seven triggers are presented as a group at the end of this section.

8. Product Diversification Approach. In the best of organizations—companies like Apple, Proctor & Gamble, and Coca Cola—product variation is the result of a carefully considered policy of product diversification, anchored to a well-defined and unified development strategy. Key to such a strategy is clarity on two foundational conditions. First, the company is clear on its long-term goals—where it wants to go, its vision of the future. Second, the company has decided on a growth strategy—how it plans to get where it wants to go, including the role new products will or will not play in getting there.

Another type of firm may not figure new products into its growth plans at all, relying instead on acquisitions, licensing, and/or geographic expansion. But when new products are named as players, a company would be wise to identify the exact types of new products, tie specific financial goals to each, and designate screening criteria each new product needs to satisfy. This is a good first step toward developing a sturdy process for introducing new products.

If your company is product-driven, you need to understand and define your product strategy—whether by choice or deed. This is policy development on the highest level. Not doing so subjects your organization and its future to forces it does not control. When that happens, diversification can trigger less benefit, if any at all, and more negative variety.

If you have not yet determined an official new product approach, consider the three following and see which best fits your company. A combination might be best. As part of that, consider wisely and well the impact of each strategy on parts and parts type proliferation and the attendant costs.

- Product Innovation
- Product Improvement, Extension, and Revision
- Reactive Development

Approach 1: Product innovation. Companies that adopt *product innovation* as a growth approach are looking for high returns and are willing to take risks. In such companies, new products can typically contribute at least 20% to annual revenues and sometimes as much as 60%.

In this kind of organization, new products are the lifeblood—and everyone knows it. Significant dollar and manpower resources are made available specifically to the effort. While the new product mix may include extensions and improvements (see below), the development process is squarely focused on the introduction of fresh, original, innovative products that propel the company into new markets or even entirely new businesses. Apple, Amazon, and Google are good examples of this type of organization.

Companies that follow this product expansion approach anticipate substantial increases in parts, part types, and the corresponding costs that innovation typically triggers. Without the guiding principles and practices of variety effectiveness, parts, parts types, and costs can needlessly swell. And, while the enterprise may be successful—even wildly so, as in the case of the companies cited above—profits may fall far short of their potential.

Approach 2: Product improvement, extension, and revision. In contrast with innovation-centered companies, organizations that adopt this improvement approach to new products want to play it safe. They want strong growth but are also interested in minimizing risks.

Advanced products may be a part of their growth strategy, but these companies concentrate on improving, extending or revising existing successful product lines. Commonly referred to as the upgrade or full-line approach, product improvement is the central stratagem of most carmakers and can also be widely seen in retail, cosmetics, clothes, and the non-prescription pharmaceutical industry. The names of these products are often good indicators of their genesis—*New Improved Tide, Miller Lite Beer, Kraft Free Dressing.*

Developing flankers is another way to expand a product line: new products that strengthen an existing market niche by differentiating it further. Pet foods with their extensive age and size group segmentation are a good example; disposable diapers and jeans are two others. Each flanker caters to a slightly differentiated customer—but all plug into the same need as the original product and do not compete directly with each other. These kinds of new product tactics generally contribute 10% or less to yearly revenues but can reach 20% or higher.

For companies that do not have a comprehensive practice of variety effectiveness in place, this product strategy is perhaps the most susceptible to exploding costs. With the multiplicity of products this strategy can trigger, there is greater risk that superfluous offerings, unnecessarily complex products, and myriad unwarranted new parts will get introduced.

Approach 3: Reactive development. This low-growth approach responds to product competition rather than creating it. The focus is on *me-too* products that display some slight variation or improvement over rival offerings. Resultant products usually account for annual revenue percentages as low as 5% or even less.

This is often the approach adopted by companies that have a high degree of security in the marketplace and perceive little threat from competitors. When and if a new product possibility appears, the organization evaluates it on an opportunistic basis. Many banks fall into this category, changing rates, for example, only in response to changes by the Federal Reserve and offering new services only after the competition has proven them successful.

Since this reactive approach tends to keep new products to a minimum, complexity and parts count triggered by new products can be kept to a minimum as well. By the same token, because such organizations do not grow their offerings pro-actively, they are likely to have few mechanisms in place to protect themselves against the pitfalls of even their circumscribed expansion (see The Eight Runaway By-Products in Chapter Two). Such companies can end up paying more dearly for their limited efforts than other, more product-aggressive firms.

9. Products as Separate Entities. Another issue related to the absence of a unifying new product development strategy occurs when products are handled as separate entities. In that case, they are developed on an individualized basis, with no seeming relationship to other products. No attempt is made to group disparate products into coherent family categories or to follow a line of logic linking them to one another, either conceptually or structurally.

Inside the company, the result is many more products and parts than would otherwise be required—plus ballooning complexity of processes and control points. To the marketplace, the company exhibits a confusing array of products that often vie with one another for customers.

10. Long Lapses between Products. There are legitimate, market-driven reasons for lapses between new product introductions as well as reasons connected to the internal availability of resources. But when new products are not anchored in a well-defined development strategy, product fragmentation occurs. The net effect is that new products often appear to be unrelated.

11. Different Designers/Different Design Concepts. Companies pay designers handsome salaries for their imagination and creativity. But some designers tend to work in isolation and, either by default or intention, promote their own individual ideas about products.

Many prefer to work as individual contributors, without concern for integrating their work into a larger concept of product line or working cooperatively with others that have a stake in effective design. Isolation between designers is intensified by the differing developmental concepts held by their managers or groups, making coordination between individual designers or groups that much more difficult.

As a result, company products as a whole do not reflect a group consistency or design coherence. Variety in products and parts can appear arbitrary and random. This haphazard effect is further aggravated if design managers come and go.

12. Cost Targeting in New Product Development. Engineers work hard to meet cost constraints handed to them by Sales and Marketing. Unfortunately, this sometimes results in new products that do not consider existing capabilities or inventory.

Typically, Sales and Marketing do not have a good understanding of the true cost of inventory and process proliferation, especially when a company's misplaced focus on L+M+O costs encourages design engineers to dismiss the use of existing parts.

In addition, a company's parts and process classification systems, even when computerized, often do a poor job of meeting the needs of design engineers. Specification information on existing parts and processes can be extremely limited—and designers often complain about the running around they have to do in order to find the data they need (also see below).

13. Technical Solutions. Companies put a host of techniques at the disposal of their engineers in the belief that these can result in improved products. They often do. If improperly applied, however—or if applied outside the context of a larger, unified product approach—they trigger negative variety. Value engineering, computer-aided design/computer-aided manufacturing (CAD/CAM), and Quality Function Deployment (QFD) are three techniques that fall into this category.

Value engineering is a technique aimed at satisfying the functional requirements of specific, individual products at the lowest possible cost and the highest value (hence the term "value" engineering). The result is a product for which functional costs have been minimized, but variety costs, across the company's product universe, can increase severely.

CAD/CAM is a powerful computerized system for integrating the product design and development cycle with the manufacturing process. Because of the ease with which CAD/CAM allows engineers to develop new products, it is easy to pay less attention to possible product carry-overs (the process of utilizing existing parts or components in new product design). Without specific policies to the contrary, product engineers usually find it "more convenient" to design from scratch. The result is that many more parts can get introduced into the system than necessary.

QFD is an effective tool for aligning product features and characteristics with a specific customer base and identifying cost tradeoffs.

While QFD can help a company develop verifiably customer-driven products, it does not specifically track the impact of the variety it generates across the company's product spectrum.

14. Product Life-Cycle Decisions. As products move through their life cycles, they reach a point when sales volumes drop off, economies of scale are reduced, and margins erode.

Engineering departments often express concern over mounting requests to value engineer products on the down side of their product life cycles. Under their own set of pressures, Sales and Marketing often badger Engineering to "get the cost out" of those same products in the hope of keeping margins high despite eroding sales.

The question, however, is: Will the processing and control costs associated with value engineering be paid for by remaining sales? In far too many cases, value engineering drives the true cost of products up—not down. Specifically, it drives up control costs and costs of stock-on-hand, however inadvertently. This is an artificial improvement—and its cost far exceeds any cost-per-piece reduction.

The lack of a coherent product development approach is almost always accompanied by weak (or non-existent) procedures for improving or discontinuing products. As a result, the company continues to invest in products that no longer command significant market share. As an operations manager in such a company once quipped: "If there is any hint of life left in the product, Sales will demand 100% support. So why should we be surprised when, a week before a product line is declared dead, a thousand parts arrive for it on the loading dock?"

As a case in point, recommendations for addressing the nine triggers described in this section (Figure 4.10) represent important changes on the policy level—but only when taken as a whole. They will be weakened if considered individually and not in association with each other. We advise you to take the time to integrate them and make them into a coherent whole, as part of your Product Development approach. In this way, the strength of their positive impact is increased.

In some companies, this may be its own kind of challenge, one that rivals the discrete improvements of those nine triggers. So be it. The cost and time savings inherent in developing coherency in your new product development/introduction function will far outweigh the upfront effort. We encourage you to pursue this and take on the challenge. Changing policy changes everything.

Figure 4.10. Improved Policies for Triggers 8 to 14: Product Development

Trigger 8 to 14
IMPROVEMENT RECOMMENDATIONS: PRODUCT DEVELOPMENT

Identify flawed product development policies that:
a) treat products as separate entities
b) isolate designers and don't provide them with a unifying development approach
c) expect engineers to meet cost targets without first understanding true cost and/or without access to accurate and complete data on existing parts and processes
d) use technical and/or computerized systems that let negative variety breed unchecked
e) use value engineering to target the artificial improvement of cost-per-piece savings in mature products

Replace them with the following improved policies.

Improved Policy (8 to 14) .1
Educate marketers and engineers in VEP principles, including true cost and the difference between effective and ineffective variety.

Improved Policy (8 to 14) .2
Favor and utilize existing parts and processes in developing new products.

Improved Policy (8 to 14) .3
Assume that existing parts and processes *are* the least-cost solution; compute these at the lowest comparable "standard" cost.

Improved Policy (8 to 14) .4
Base cost targeting on: a) number of parts in the product; and b) inventory turns of parts in the product. Avoid L+M+O.

Improved Policy (8 to 14) .5
Arrange all product lines in categories along a projected life-cycle curve.

Improved Policy (8 to 14) .6
Develop procedures for controlling or minimizing the introduction of new parts caused by the use of technical and/or computerized resources and without a product-wide perspective.

Improved Policy (8 to 14) .7
Develop criteria for qualifying a product for value engineering based on its lifecycle position (e.g., those in the latter third of the curve). Do this for all products except where an outright product defect is present.

Improved Policy (8 to 14) .8
Do not make customer-specific products that have no economies of scale candidates for cost reduction except where a part or process must be totally removed.

Improved Policy (8 to 14) .9
Develop a costing system that allocates overhead based on: a) number of parts in a product; and b) inventory turns per part (see Trigger 3 above for further discussion).

D. Negative Triggers in IT/Data Systems Policies

Your data systems policies are a hotbed of triggers for negative variety. Though it is the worst hard work you can ever imagine, you must address these policy triggers of negative variety or they will run rampant and never be brought into balance in your company. We will now look at the following three policy triggers, under a single heading: Product Documentation, Classification Systems, and Computer Support.

15. Product Documentation, Classification Systems, and Computer Support. Document and data maintenance is a challenge in any business. Coupled with manually-maintained drawings, simple changes often result in the revision of literally hundreds of files and routings and a never-ending backlog. Companies habitually turn to IT for solutions. But are computer-based solutions problem-free? Let's take a look.

The enterprise often invests in computer-aided design (CAD) in the hope of achieving flexibility and organization for systematic part and process classification and routings, drawings and their storage and revisions, and other data-driven tasks. But these activities are rarely undertaken within a unifying data framework.

As a result, classifying new product or parts descriptions and assigning part numbers are seldom governed by: a) a standardized nomenclature; b) a standardized description format; or c) steps to ensure that part numbers carry the right computer codes and the proper code-number sequencing. Additionally, specification information on existing parts and processes is often inaccurate, incomplete or incorrectly entered— or a combination of all three. A challenge of gargantuan proportions begins to take shape.

For example, engineers need to access detailed attribute characteristics in order to decide if an existing part is sufficient for a new design. If these data are not complete, correct, and easily accessible, the engineer may well decide to design from scratch. Another victory for negative variety!

These shortcomings in a company's data approach produce early and lingering failures in product development that impact many other functions. Design engineers are left with the brunt of the confusion. They wrestle, for example, with part descriptions and part number sequences that appear identical but are only similar. No wonder they so stridently complain about having to chase down data on parts and processes.

It is not hard to understand how this spills over to problems with the construction of Bills of Material (BOMs) and the storage of parts. Sales and Marketing also bear the burden of the mayhem triggered in the data entry process as they sort through multiple SKUs for the same product as well as multiple shelf names for distributors. For more, see Chapter Seven.

Figure 4.11. Improved Policies for Trigger 15: IT and Data Systems

Trigger 15 IMPROVEMENT RECOMMENDATIONS: IT/DATA SYSTEMS
Identify flawed data policies that: a) do not provide engineers with the detail they need to make effective design decisions b) do not provide for systematic additions and changes to the database **Replace them with the following improved policies.**
Improved Policy 15.1 Do all drawings using CAD or photo-capture technology.
Improved Policy 15.2 Establish criteria for exempting certain changes to old drawings.
Improved Policy 15.3 Develop attribute templates and a database system—maintained by the engineering function—that support designers in making variety effective decisions on parts and processes.
Improved Policy 15.4 Make the improved database developed under Improvement Policy 3 available to all employees.
Improved Policy 15.5 Have the IT department take a lead role in revising the computer model to more accurately reflect organizational and policy changes that support variety effectiveness.

E. Negative Triggers in Operations Policies

Once the product is designed, launched, and in operations, the two main trigger areas of negative variety are:

    16. Lot Sizing

    17. Capital Equipment and Process Improvement Justification

16. Lot Sizing. Unwritten policy on lot sizing results in overproduction and bulging inventories. Economic order quantity (EOQ) assumes that long setup time and defects are unavoidable and therefore justifies large lot production. Despite impressive advances in quick equipment changeovers and fail-safe devices *(poka-yoke)*, many organizations continue to accept large lots as a given.

Even when some effort is made to increase quality and reduce setup times, management can have a tough time selling supervisors, production planners, and hourly employees on the idea that inventory is not a saving grace but a costly waste.

Instead, they cite long setup times and loss due to defects as the reasons for building excess stock and maintaining buffer quantities. In a similar way, purchase-order release quantities often reflect a "more-is-better" attitude. On the one hand, companies adopt an informal cost-per-piece policy in purchasing parts; on the other, suppliers offer significant price breaks on large-order quantities. In addition, while Enterprise Resource Planning (ERP) systems may be in place and monitored faithfully, planners often increase the suggested ERP quantities—just-in-case—before handing off requisitions to Purchasing. Purchasing, in turn, factors in "a little extra" for problems in the supplier's process or a supplier's inability to deliver small lots.

The end result is a lot of extra quantity on hand, ballooning storage needs, an increase in the number of core and supporting processes (control points)—and low inventory turns in all parts, purchased and made (sometimes as low as three a year).

Figure 4.12. **Improved Policies for Trigger 16: Lot Sizing**

Trigger 16 IMPROVEMENT RECOMMENDATIONS: LOT SIZING
Identify flawed policies relative to lot sizing that: a) promote large lot production and decision-making based on EOQ or other wobbly formulas b) valuate inventory as an asset c) seek price breaks from suppliers for ordering parts in large quantities **Replace them with the following improved policies.**
Improved Policy 16.1 Institute lot-size-of-one as the ideal for all manufactured parts and end-items, even if you believe your production model will not permit it.
Improved Policy 16.2 Purchase parts in quantities of no more than one-quarter stock-on-hand, regardless of the cost per part.

17. Capital Equipment and Process Improvement Justification. Many companies use the yardsticks of *efficiency* and *utilization* to justify investing in capital equipment and asset expansion. In isolation, however, efficiency can sub-optimize a process because it does not factor in actual market need. Similarly, utilization as a measure discourages frequent machine changeover—which, once again, serves to generate overproduction and excessive storage needs. Both measures are outdated and can impede variety effectiveness because they impede the reduction of parts and processes. In addition, management often selects equipment based on personal preferences (equipment pets) rather than sound logic.

Make sure to consider these scenarios as you flesh out your revised VEP-driven policies on equipment purchasing and process improvement.

Figure 4.13. Improved Policies for Trigger 17: Purchase Justification

Trigger 17
IMPROVEMENT RECOMMENDATIONS: CAPITAL EQUIPMENT AND PROCESS IMPROVEMENT JUSTIFICATION

Identify flawed policies that allow equipment purchases and process improvement expenditures to be based on:
a) cost-per-part consideration
b) large lot production
c) machine efficiency (overproduction)
d) equipment pets
Replace them with the following improved policies.

Improved Policy 17.1
Do not use efficiency and utilization calculations or cost-per-part and machine efficiency considerations in isolation for justifying process improvements or investing in equipment.

Improved Policy 17.2
Establish a documented procedure for qualifying an investment to improve a process and/or purchase equipment.

Stem The Tide

The causes of negative variety in an organization are myriad but not countless. Furthermore, they can be tracked. And when you do track them, you will see that they are invariably linked to policies. Whether formal or informal, the power of policy cannot be over-estimated and therefore must be brought under enterprise scrutiny and control.

The first step is to surface and formalize your policies. Kept informal, they are dangerous, held in place by habit and a tacit, damaging agreement that the status quo must not be questioned. Every part of the organization makes its own silent unhappy contribution to this: the continuation of "things as they are." But when educated in the VEP process, those same departments become self-reflective and positioned to make a positive contribution to the goal of variety effectiveness.

There's no getting around it: When you ship your product, you ship your corporate decisions—your policies and practices. If your product strategy, for example, does not consider the impact of new parts on the entire organization relative to variety effectiveness, it is bound to show in your product—in its quality, performance, cost to you, and/or price to your customer. Everyone will pay extra for it, including you.

Begin with an active search for the formal and informal policies that trigger unwarranted variation. This fact-finding mission is what this chapter is designed to help you organize. No department, person or policy is exempt. Once a suspect policy is spotted, your job is to ask and answer a single driving question: What is the impact of this policy on our effort to reduce or prevent unwarranted variety? Then trace the causes back to their roots—and eliminate them, inadvertent as they may be.

Rather than allow policy to distort and obstruct the effort of your company to grow, expand, and flourish in today's competitive markets, change those policies to support effective variety. Begin to stem the tide of unrestrained variation and redirect it toward higher profitability.

See Figure 4.14 for the full set of Policy Triggers of Negative Variety discussed in this chapter and the associated recommendations for improving them. In Chapter Five, we consider a set of broad-level guidelines to help us design products for profit.

Figure 4.14. **Summary: All Policy Triggers and Recommendations**

SUMMARY: ALL POLICY TRIGGERS AND RECOMMENDATIONS	
Trigger 1: Supplier Base Selection. Identify flawed supplier policies that make supplier selections based on: a) cost-per-piece only b) purchasing pets c) calamity protection **Replace them with the following improved policies.**	
Improved Policy 1.1	Establish partnerships with best suppliers and provide 90% of the business for a particular commodity to the primary supplier that can best serve the company's requirements for service, quality, price, and delivery.

	SUMMARY: ALL POLICY TRIGGERS AND RECOMMENDATIONS (continued)
Improved Policy 1.2	Set a target for supplier reduction that is aggressive enough to require a genuine change in policy (for example, reduce supplier base by 40% in the next year).
Improved Policy 1.3	Do not add a new supplier to the supplier base without directly offsetting another supplier that provides the same commodity.
Trigger 2: Make vs. Buy Decisions. Identify flawed procurement policies that result in automatic parts out-sourcing due to: a) cost-per-piece accounting b) restricted internal capacity c) limited process capability d) "just not feeling like making it in-house" **Replace them with the following improved policies.**	
Improved Policy 2.1	Design parts to favor existing or planned in-house part production process capability.
Improved Policy 2.2	Where a choice of raw materials exists, design parts to standardize on the most robust material that satisfies function across many products.
Improved Policy 2.3	Let raw material cost, capacity, and process capability be the only constraints in determining make vs. buy decisions. Do not consider L+M+O overhead costing.
Trigger 3: Overhead Cost Allocation. Identify flawed accounting policies that allocate product and inventory costs based on an obsolete Labor + Material + Overhead (L+M+O) formula. **Replace them with the following improved policies.**	
Improved Policy 3.1	Develop a costing system that allocates overhead based on: (a) number of parts in a product, and (b) inventory turns per part.
Improved Policy 3.2	Do not use a direct labor-based overhead cost structure for determining inventory costing, except in valuating it for taxes.
Trigger 4: Response to Customer Requests. Identify flawed customer-response policies that: a) give customers whatever they ask for b) provide indefinite product support c) obsolete a product only when internal complaints get loud enough d) obsolete a product, arbitrarily, in response to periods of product non-activity. **Replace them with these improved policies.**	

	SUMMARY: ALL POLICY TRIGGERS AND RECOMMENDATIONS (continued)
Improved Policy 4.1	Determine product obsolescence, based on the tracking of product life cycle.
Improved Policy 4.2	Set up a continuous product profitability feedback loop that defines and identifies trigger points for retiring unprofitable products.
Improved Policy 4.3	Determine and publish an obsolescence timetable (for example, three-year notice before obsolescence; five years for maintenance and repair).
Improved Policy 4.4	Provide quotes for customer-specific products that state minimum annual usage amounts required for continued company product support.
Improved Policy 4.5	Provide all sales and marketing personnel with on-going education on effective vs. ineffective variety.
Improved Policy 4.6	Implement these improved policies as an on-going and required part of the marketing function.

Trigger 5: Margins. Product Pricing, and Discounting. Identify flawed pricing policies for setting price based on:
a) profit margin as calculated by the traditional L+M+O formula
b) what the market will bear
c) economic order quantity
Replace them with the following improved policies.

Improved Policy 5.1	Base profitability on the comparison of net selling price to total cost, in which material and overhead are allocated on a part/product basis.
Improved Policy 5.2	Hold prices of mature products constant in mature markets.
Improved Policy 5.3	Figure product price based on market forces—not L+M+O.
Improved Policy 5.4	Provide product discounts based on total order quantity, without regard for minimum releases (EOQ).

Trigger 6: New Product Market Requirements. Identify flawed new product policies that result in the introduction of:
a) products into new markets without sufficient research
b) so-called "new" products that cannibalize existing products or are simply revisions of them
Replace them with the following improved policies.

SUMMARY: ALL POLICY TRIGGERS AND RECOMMENDATIONS (continued)	
Improved Policy 6.1	Develop a well-defined new product development strategy that includes screening criteria for evaluating new product requests.
Improved Policy 6.2	Develop new products that fill voids within previously targeted markets instead of products that are redundant.
Improved Policy 6.3	Sell existing products to any market but do not undertake new product development for markets not specifically targeted in your strategic plan.
Trigger 7: Cost Targeting and Cost Reductions. Identify flawed cost-targeting policies that trigger new products based solely on a traditional lowest-cost profile and not on a true-cost profile. **Replace them with the following improved policies.**	
Improved Policy 7.1	Base cost targeting on two factors: 1) number of parts in the product; and 2) number of inventory turns of parts in the product (with an annual turn of four equaling unity).
Improved Policy 7.2	Avoid entering markets solely on the basis of a lowest-cost product.
Trigger 8-14: Product Development. Identify flawed product development policies that: a) treat products as separate entities b) isolate designers and don't provide them with a unifying development approach c) expect engineers to meet cost targets without first understanding true cost and/or without access to accurate and complete data on existing parts and processes d) use technical and/or computerized systems that let negative variety breed unchecked e) use value engineering to target the artificial improvement of cost-per-piece savings in mature products **Replace them with the following improved policies.**	
Improved Policy (8 to 14) .1	Educate marketers and engineers in VEP principles, including true cost and the difference between effective and ineffective variety.
Improved Policy (8 to 14) .2	Favor and utilize existing parts and processes in developing new products (shared vs. unique).
Improved Policy (8 to 14) .3	Assume that existing parts and processes *are* the least-cost solution; compute these at the lowest comparable "standard" cost.
Improved Policy (8 to 14) .4	Base cost targeting on: a) number of parts in the product; and b) inventory turns of parts in the product. Avoid L+M+O.

SUMMARY: ALL POLICY TRIGGERS AND RECOMMENDATIONS (continued)	
Improved Policy (8 to 14) .5	Arrange all product lines in categories along a projected life-cycle curve.
Improved Policy (8 to 14) .6	Develop procedures for controlling or minimizing the introduction of new parts caused by the use of technical and/or computerized resources and without a product-wide perspective.
Improved Policy (8 to 14) .7	Develop criteria for qualifying a product for value engineering based on its lifecycle position (e.g., those in the latter third of the curve). Do this for all products except where an outright product defect is present.
Improved Policy (8 to 14) .8	Do not make customer-specific products that have no economies of scale candidates for cost reduction except where a part or process must be totally removed.
Improved Policy (8 to 14) .9	Develop a costing system that allocates overhead based on: a) number of parts in a product; and b) inventory turns per part (see Trigger 3 above for further discussion).

Trigger 15: IT/Data Systems. Identify flawed data policies that:
a) do not provide engineers with the detail they need to make effective design decisions
b) do not provide for systematic additions and changes to the database
Replace them with the following improved policies.

Improved Policy 15.1	Do all drawings using CAD or photo-capture technology.
Improved Policy 15.2	Establish criteria for exempting certain changes to old drawings.
Improved Policy 15.3	Develop attribute templates and a database system—maintained by the engineering function—that support designers in making variety effective decisions on parts and processes.
Improved Policy 15.4	Make the improved database developed under Improvement Policy 3 available to all employees.
Improved Policy 15.5	Have the IT department take a lead role in revising the computer model to more accurately reflect organizational and policy changes that support variety effectiveness.

SUMMARY: ALL POLICY TRIGGERS AND RECOMMENDATIONS (continued)

Trigger 16: Lot Sizing. Identify flawed policies relative to lot sizing that:
a) promote large lot production and decision-making based on EOQ or other wobbly formulas
b) valuate inventory as an asset
c) seek price breaks from suppliers for ordering parts in large quantities
Replace them with the following improved policies.

Improved Policy 16.1	Institute lot-size-of-one as the ideal for all manufactured parts and end-items, even if you believe your production model will not permit it.
Improved Policy 16.2	Purchase parts in quantities of no more than one-quarter stock-on-hand, regardless of the cost per part.

Trigger 17: Capital Equipment and Process Improvement Justification. Identify flawed policies that allow equipment purchases and process improvement expenditures to be based on:
a) cost-per-part consideration
b) large lot production
c) machine efficiency (overproduction)
d) equipment pets
Replace them with the following improved policies.

Improved Policy 17.1	Do not use efficiency and utilization calculations or cost-per-part and machine efficiency considerations in isolation for justifying process improvements or investing in equipment.
Improved Policy 17.2	Establish a documented procedure for qualifying an investment to improve a process and/or purchase equipment.

Trigger 8-14: Product Development. Identify flawed product development policies that:
a) treat products as separate entities
b) isolate designers and don't provide them with a unifying development approach
c) expect engineers to meet cost targets without first understanding true cost and/or without access to accurate and complete data on existing parts and processes
d) use technical and/or computerized systems that let negative variety breed unchecked
e) use value engineering to target the artificial improvement of cost-per-piece savings in mature products
Replace them with the following improved policies.

Improved Policy 8-14.1	Educate marketers and engineers in VEP principles, including true cost and the difference between effective and ineffective variety.

SUMMARY: ALL POLICY TRIGGERS AND RECOMMENDATIONS (concluded)	
Improved Policy 8-14.2	Favor and utilize existing parts and processes in developing new products (shared vs. unique).
Improved Policy 8-14.3	Assume that existing parts and processes *are* the least-cost solution; compute these at the lowest comparable "standard" cost.
Improved Policy 8-14.4	Base cost targeting on: a) number of parts in the product; and b) inventory turns of parts in the product. Avoid L+M+O.

Smart Simple Design/Reloaded

CHAPTER 5

Hot Products: Design For Overall Cost

A hit product goes beyond just selling well. A true hit product establishes an entirely new franchise in the marketplace. You can design these products, and they not only win awards but sell like crazy.

— *BOB BRUNNER, Industrial Design/Apple Computer Inc.*

Leveraging the power of design is the name of the game today. Product design is a core competency that smart companies are using to drive their entire product development process. The prize? Hit products, billion-dollar sellers, and award-winning designs such as: Apple's iPad and iPhone, Gillette's Fusion ProGuide Power Razor, the Nike Free Flyknit and Free Hyperfeel footwear, and Tesla Motors electric cars. These design-driven products are transcending the traditional norms of market success. They are defining whole new product categories.

We begin by defining terms. *Industrial designers* are responsible for conceptualizing and pre-inventing all aspects of a new product—its function, geometry, style, and visual impact. Sometimes called stylists, industrial designers understand the needs of the customer and create the total product with that in mind. Their job is to shape a vision of the product that works aesthetically and functionally, often creating results that expand market boundaries.

Product engineers are inventors and validators. They find ways to realize the form, fit, and function of the product concept the designers generate. Product engineers make things work; they solve problems. They take what designers conceptualize and determine the precise geometry (geometric functionality) required to make the product perform in real time and real space, and all within the cost constraints of their product developmental budget. Part of their challenge is to preserve in the final product as many of the original design elements as possible—the appeal that the company spent big design dollars to create.

It is inaccurate—though all too easy—to say that designers focus on the outside of a product and engineers focus on the inside. Even when housed in different departments, the two groups work in tandem. When co-located in teams (and that's the growing trend), engineers help design and designers help engineer. To invent and create—that is why they both went to school in the first place: to get a license to devise hot new products. Working together, industrial design and product engineering get new technology out of the lab and into the hands of the buying public.

Our discussion of product design and variety effectiveness begins, once again, at the computer.

CAD: A Double-Edged Sword

Geez, I must be more productive. I'm on CAD!
— *EVERYONE*

When CAD came on the scene decades ago, it revolutionized the product development process. Lead times were slashed, product features reached new levels of imagination, and the differentiation between products became increasingly subtle. Companies that invested in CAD (and CAD-CAM) enjoyed tremendous success and short payback periods. They provided their designers and engineers with a tool that gave them a wondrous new window on innovation as a competitive advantage. Plus, CAD linked with CIM (computer-integrated manufacturing)—and then later with Rapid Prototyping and 3D printing—took so much of the pain out of making and delivering new products.

Since all these techniques use CAD functions as a base, they are collectively referred to in our discussion as CAD.

Over-Designing: Overusing Your Strengths

Yes, CAD offers companies a vast new creative capability, allowing engineers to develop innovative, highly differentiated products at a fast clip. While a breakthrough technology of enormous significance, CAD has strengths that can be and are overused. CAD is a double-edged sword.

Driven by a nearly universal call for market-in design, every new and existing product seems subject to upgrade and enhancement. Distinctive products—and distinctions between products—have become an obsession. In the search for new products, many companies send designers out in droves to be with the consuming public and discover opportunities for new hot products. Armed with CAD, lean production, workplace visuality, QFD, and other world-class methods, manufacturers are confident that they can provide customers with everything they want—and at an appealing price. The market, already highly segmented, has fragmented into thousands of slightly differentiated bits. We repeat: The age of mass customization has arrived.

The computer, and the canny software that supports it, allows businesses to generate huge annual crops of new products with levels of speed, ease, and ingenuity previously unthinkable. The same computers provide organizations with the capability of storing and manipulating the immense amounts of information needed to support the design, development, marketing, production, and sale of these products. But these accelerating computer capabilities were never designed to manage or regulate themselves—or to put a lid on their prodigious output.

Revising The Mindset: Design For Overall Cost

At the beginning, CAD provided designers and product engineers with a free hand to develop new products. But the playing field has since shifted again. First, product life cycles collapsed. Time-to-market is a fraction of what it used to be. Today's competitive edge evaporates as access to new technologies widens. In the United States, new safety and emissions requirements are taking a big bite out of margins. While companies won't risk safety, reliability or performance, to stay even they also have to find new opportunities to cut costs.

It is time to revise the product design and engineering mindset. In Chapter Four, we recommended that companies review and, where needed, revise their policies to better support the principles of effective variety. As part of that, we proposed a set of specific policy improvements related to product development.

In the remainder of this chapter, we discuss two additional dimensions of revising the design and engineering mindset. The first identifies four proposals aimed at reducing overall cost and helping designers minimize negative variety. The second dimension looks at barriers to adopting these proposals. Our four proposals for designing for overall cost are:

1. Design from the outside in
2. Know when average is good enough
3. Use fewer parts—use shared parts
4. Get sales involved from the get-go

Design from the Outside In—Put Value Near the User

The old two-step of product development used to run something like this: one, the company stumbles upon a great new technology; two, the company finds some place (or some way) to sell it and designs a product around that. The product got designed from the inside out.

More and more, that is no longer working. Today's hot products begin from the *outside in*. That's been Apple's approach from the beginning: design from the outside in—from the point of view of the user. Engineers at Apple discovered, for example, that people didn't really want small computers *per se;* they wanted mobile computers. Size was just one dimension of that. That led Apple to a whole new set of features and to award-winning products that never seem to end—and all of them sell wildly.

Nike faces that same challenge with every new model season. As the head of Nike's sneaker lab once said, "It's getting harder and harder to build a better sneaker—and even harder to come up with innovations that people can easily see."

This was the core—and the magic—of Chrysler's approach to designing the Neon, its compact launched in 1994. Is this example too out-of-date for your consideration? Not at all. The principles remain relevant.

In 1994, the challenge for Chrsyler was not just designing a great new compact; that would have been daunting enough. The challenge was to do it at a cost that let the sticker price—$8,995—be a major competitive feature and beat the Japanese at their own game. But that wasn't all. Chrysler also wanted the Neon to be first-rate in performance and low-cost to make. And, while we're at it, said the boss, bring this baby in at, or under, a $1.3 billion budget—and inside an accelerated time-to-market window of 31 months. Those last two items were big ones, considering that, during the same period, Ford needed $2 billion and 5 years to develop its latest Escort, and GM required $5 billion and 7 years to launch the Saturn.

Where to begin? Early on, the team (a cross-functional group Chrysler calls a "platform team") set its priority: Design the car from the outside in. Put the money where it really counts. Don't scrimp on features, performance or safety; but cut costs where the customer won't notice or won't care. They sang the variety effectiveness song.

According to *Business Week,* small-car owners told Chrysler researchers that power windows were not important to them so the team chose the crank variety for the back seat. Owners also saw nothing special in a four-speed automatic transmission, even though most competitors were offering them at that time. So Chrysler engineers adapted one of their existing three-speeds, saving the company more than $300 million dollars in developmental, tooling, and other operational start-up costs.

The idea is to spend money where it really counts. Again, Chrysler discovered that small-car owners worry a lot about safety. So team members bit the bullet and made Neon the first subcompact with dual air bags standard; that feature comprised a full 10% of the car's total parts bill. In its turn, supplier TRW Inc. lowered that cost substantially by designing a single, cheaper impact sensor for the air bag—to replace the usual three. In addition, Neon had reinforced doors three years before tough federal side-impact standards were put in place. That's variety effectiveness. Value is in what gets used—not what gets built.

It would be a mistake to think that the strategy Chrysler used in designing the Neon should be restricted to low-end products. Here is a story about the other end of the product spectrum, told to us by a VP of Design at a successful automotive supplier.

"In its heyday, a few of us went to Lotus cars to see its short-run production system. I grew up a car nut and Lotus was always one of my favorites. So we get there and I'm looking inside these cars and I see— plywood and fiberglass! You would have thought they were building boats, not $75,000 cars! And yet, as they added the layers in, you saw that they put all their money where the end-user was. The leather, knobs, sound system, carpets, glass, and all the other things the customer saw was just the best! Two inches away was this stuff called 'plywood.' I walked away thinking, 'Whew. What a cheap car underneath!' And then I said, 'Boy, is that smart!'"

Know When Average Is Good Enough

Companies are always challenged to delight the user. Often this is taken to mean: Build customer loyalty by delivering more than the customer expects. A great concept *if* you can afford it. In Japan, at a time when manufacturers faced a sharply stronger yen, companies scrambled to cut costs. The result was a major shift in thinking for Japan and a realignment of priorities.

Honda, for example, succeeded in cutting costs on one of its minicars—sold exclusively in Japan—without losing features that appeal to the car's mostly female owners. First, Honda substituted a regular trunk for the more expensive hatchback in an earlier version and racked up considerable cost savings due to the less expensive superstructure required by a trunk. These and other efficiencies (see below) let Honda cut the price of the model, add an air bag, and still post sales 21% above target in the first selling quarter.

The facts are plain: Companies are shifting away from insisting on perfection and are looking for areas where average will sell. For example, some carmakers are questioning whether to continue spending big bucks to ensure particularly tight seams between body panels for the U.S. market where buyers don't seem to notice. If the market accepts lower standards, is it really a good idea to insist on higher ones?

To do this, product designers and engineers need to have their priorities straight. As mentioned, they need to design the product from the consumer's viewpoint, adding a delightful surprise or two but also making sure the product is not overdesigned. That means the insides of the product need to function properly—but they do not need to make a fashion statement.

Use Fewer Parts—Use Shared Parts

The third element of the revised mindset is: Use *fewer* parts and make sure as many of those as possible are *shared*. We have already advocated rating products on the basis of their unique-to-shared parts ratio. The absolute goal is: as many shared parts as possible. Put the value where the customer can see it. Cut costs where it is invisible to the user.

In today's disinflationary marketplace, consumers will buy a product strictly on the basis of low price—as long as its quality, performance, and value are comparable to its rivals! Low-cost product development requires companies to practice *design-to-price* or target-pricing: setting the price based on an acceptable profit margin and backing out cost targets from there.

One well-known computer company put this practice to good use after years of battering by its low-cost competitors. It came out with computers that cost up to 60% less than its competitors. How? Using target-pricing as a guide, engineers designed products with fewer parts and reused parts from existing designs to achieve cost targets. It was as simple as that—and as smart. The results were amazing. The first products manufactured under the new target-pricing system were out in less than eight months; less than a year after launch, sales volume had skyrocketed by 64% and profits nearly doubled.

Milacron Inc., a machine tool manufacturer based in Cincinnati, is another company with the same strategy. When Milacron could not raise prices in the face of disinflationary pressures, it learned to build its machine tools with 30% to 40% fewer parts. On one of its main machining centers, it reduced the number of fasteners from 2,542 to 709 and cut assembly time from 1,800 to 700 hours. The selling price is the same—but production costs decreased by 36%. Plus, the new streamlined design takes up 60% less floor space, can be installed in two days instead of two weeks, and makes much more rapid changeovers, which sharply increases productivity. Who doesn't win from this hot new product?

In the redesign of Honda's mini-car (mentioned above), Honda engineers succeeded in reusing 40% of the parts from an earlier version. The automaker learned to share parts across product lines as well. A Honda Civic sold exclusively in Japan holds 60% of its parts in common with other Hondas. And there's more. By using fewer and simpler parts, Honda also managed to shave $700 in costs from its USA Civic.

Nissan faced—and met—the same challenge when it renounced the quest to build cars in ever more sizes, colors, and functions to satisfy the presumed whims of the world's drivers. With negative variety choking off profit margins, Nissan ordered its designers to slash the number of unique parts across all vehicles by 40%.

Get Sales Involved—From the Get-Go

Sales is the first interface with the customer, the first point of contact where the need and the opportunity for new products are identified. This is the time and the place to champion the practices described above: at the earliest possible phase of the product development process—when customers register their needs.

Here's a scenario: You're the top salesperson for a company that designs cab interiors for trailer rigs. After years of hard work, your company is designated the global source supplier for McTrucks, your biggest client, with plants all over the world. You are meeting with the client to discuss a possible new console product. Three hours later, you return with a stack of requirements which you take directly to your design team to spec out. Within an hour they contact you, saying: "There is absolutely no margin in the product." They don't really want the job. But you and they both know that you are going to have to bite the bullet and take it so your company can hold on to its global-source status. You can't afford to not do it.

Now let's imagine you as top salesperson in a slightly different scenario. Same client, same meeting. You go in, pick up the requirements for the new console, and then meet with your design team. The team scans the requirements and barks, "Whoa! No margin!" A SWAT team is immediately dispatched to search their company's product universe for common elements that might be pulled out and applied to the console. In the end, a number of existing components and structures are identified that could keep engineering work to a minimum. When the team rethinks the project, it is clear that the cost savings could be substantial and the margin definitely worth it. You call the client with the good news, mentioning that, if it all works out, you can even sweeten the deal by passing a percentage of the cost savings on to them. Customers get what they want—*and* your company gives them a product it can afford to make. Everybody wins.

An even better choice, one that we recommend, is this: Take the SWAT team with you to the client meeting. Now that's "at the get-go"! Considering the speed of change and the turbulence of the market, your sales force can be a *principal partner* in the quest for variety effectiveness. Create a team of sales and design specialists, educate them in VEP principles and practices—and send them out together for the initial consultation with clients.

Implement these four proposals to dramatically increase the likelihood that your clients will be drawn to design solutions that are variety effective and new products that are not going to bury the company. This goes a long way to ensure that your honey-of-a-deal doesn't beat you up in the end.

Barriers To Moving Forward

When properly supported, these four proposals can be instrumental in taking big chunks out of parts inventories and designing products with greater levels of positive effective variety. But there are barriers to implementing them. Here is a story where the right thing *almost* happened *several* times—but barriers got in the way.

Sid is a young designer at PUI. About a month ago, he joined a new product launch team. At last week's meeting, team members realized that if they were to stay on track, they would have to collapse the design phase radically—from eight to four weeks. They were ready to cut any corner necessary. The burning question was where and how.

The use of shared or carry-over parts came up almost immediately. Sid was asked to check out a particular latch (44R) as a possible candidate. His job was to dig up the specs from the database, grab the actual part(s), formulate some recommendations, and bring them all to the next meeting.

When Sid called up 44R on CAD, seven latches turned up—but only one had full specs and some potential. That was two days ago. Since then Sid searched for specifications on the other six; but they were either not in the database or the data were incomplete. Not wanting to give up, Sid launched a search for the drawings themselves, all the while wondering why they were not available on CAD. After another three hours of drawer pulling and paper sifting, he ascertained the following: Two of the 44R latches had been obsoleted a year ago (but part numbers had not been pulled) and two more were entirely too large. That left three with possibilities—if the product geometry could be slightly adjusted.

The meeting was just starting when Sid slid into his chair and put various drawings on the table, along with his completed report. The discussion quickly moved to carry-over parts and Sid was asked what he had come up with. Sliding three parts across the table for the team to inspect, he began to talk about shifting the geometry. By the middle of Sid's second sentence, the three parts had already completed the circuit. The team leader interrupted, saying, "Thanks, Sid. Great job! But none of these will work." It was over in a flash. There was no VEP mindset.

Without a VEP-educated mindset, the excuses multiply. Designers and product engineers like those in Sid's scenario try to make a strong case *against* looking for existing parts and *in favor* of developing 2-D and 3-D geometries from scratch. After all, not looking is certainly quicker than wading through cabinets or CAD files in search of specifications, and then still needing to scrutinize the product geometry for a fit.

Even when engineers have a complete set of CAD drawings at their disposal, they prefer not to look. The reason is easy to spot: When you try to use what someone else created, you first have to figure out how it is meant to work. Only then can you determine whether you can use it. "Well, let's see," says the well-meaning engineer, "how does this work? Ah, heck. I'm not going to fool with it." Besides, it is a bother that many times does not pay off. The engineer ends up having to create a new part after all. Easier to create it new to begin with, this logic goes.

Still, lots of designers and engineers use carry-over parts—optionally and only from time to time. Gerrie is at her desk working on a design element for a new product—a cover. Remembering a previous cover design, she pulls up the specs and reflects: "Hey, I'll bet this will work. If I carry over this existing cover, we're not going to have to prove it out or do a feasibility study. We can just use it. It's a given. Now I can spend more time in surface development—in how great I'm gonna make my new product look!"

In the two illustrations above, organized efforts to promote positive variety happen by default and not as the result of established policy. In the absence of such policy, product development decisions in support of effective variety are left to chance. Many companies have so many areas of deficiency in their new product development approach that it becomes a coin toss whether negative or positive variety ends up in the lead.

Six Barriers to Effective Variety

The discussion above illustrates some of the many barriers facing a company wanting to reap the substantial rewards of implementing VEP. We summarize six of them here that focus squarely on the Product Design function. They are:

1) no formal policy
2) limited database capabilities
3) cost as an accounting worry
4) no supporting measures and rewards
5) inadequate training, and
6) no unified strategy.

1. No Formal Policy. In the absence of formal policies supporting variety effectiveness, the product development process can only be driven by individual preference and will, instead of by the corporate intent. Without explicit VEP policies, design economies are haphazard at best—along with reductions in product costs, complexity, parts count, and development time. By contrast, with VEP policies in place, your technical staff can make concerted efforts to:

a) avoid overdesign
b) share parts and assemblies
c) design from the outside in, and
d) steer clients in the direction of effective variety.

2. Limited Database Capabilities. The company's computer database can be instrumental in the overall reduction of part inventories and designing new products with fewer and simpler parts. Or, it can be the unsuspecting cause of soaring inventories when data are not accurate, not uniform, incomplete or not easy to access. In VEP, the company's computerized database becomes a dynamic tool in design decision-making and a linchpin in launching products that represent effective variety—and not its enemy.

3. Cost as an Accounting Worry. Designers and engineers can mistakenly consider cost primarily an accounting issue. When they do, they segregate cost into a remote silo of function, separate and apart from new product activities. It is not unusual to hear managers say in effect: "Designers know what the product should look like, how it should work and perform—but not what it should cost. That's what accounting is for. Let designers stick to their knitting." Until recently, the norm seemed to be: The higher the caliber of design team, the less likely they would factor cost into their work. The same could be said for product engineers further down the line: "Cost is an accounting worry."

In these days of cross-functional product launch teams, this kind of cost mindset is more the exception than the rule—but an exception that costs dearly. With an eye on *total* cost, we need to recognize *all* cost drivers.

4. No Supporting Measures and Rewards. Product development is the creative center of many companies; yet these same companies often hesitate to put measures in place for the product design function. Instead, they consider product innovation and creative design as off-limits, not core competencies that can be measured and scored. Product design and engineering rarely use indicators to track product complexity, overdesign, overall cost, unique vs. shared-parts ratio, internal vs. external value, and the like. Similarly, a company may not use the considerable clout of its existing reward structures to recognize—and therefore to promote—practices that build effective variety.

5. No Training. Closely linked to the absence of formal policy and supporting measures is the lack of training in the principles and practices of effective variety. Most employees do their best to make positive contributions at work; but many find themselves at a loss to define what "positive" means in their specific jobs. Opportunities to increase positive variety and/or minimize the negative happen by chance—from time to time. This accidental approach can never be enough to forge an authentic strategic advantage for the enterprise.

6. No Unified Approach. All of the above add up to the greatest barrier to effective variety in the product development arena: the lack of a unified product design and development strategy. Many companies have operated for decades with no such strategy. Others think they actually have one in the single sheet of paper that lists their new product plans for the year.

A true unified approach defines and pulls together a multitude of critical factors related to the why and how of introducing successful new products. In addition to defining the overall growth direction of products and specific objectives for each entry, a unified strategy also details management's resource commitment to new products—along with the criteria, process, and people framework by which products will be developed and commercialized.

Even with a bona-fide, well-considered strategy, 100% success is never assured. Developing new products without a unified approach is like asking designers and product engineers to hit a bull's-eye with blindfolds on. Or as a manager once quipped, "Going into the design department of our company in pursuit of profitable products is like looking for cattle in a butcher shop. They are there but in a rather peculiar form."

A New Role To Play

New products are the lifeblood of every company. Designers and product engineers are constantly juggling a host of competing priorities to make great products (and services) happen—product geometry, ease of use, life-cycle costs, cost constraints, innovative design, packaging, and launch deadlines, to name a few. Under the VEP umbrella, engineers are asked to play a new role—as champions of the riches of the imagination and the economics of effective variety.

In a VEP enterprise, engineers know that effective variety means a balance point between positive (customer-driven) and negative (internally-triggered) variety. Balancing the two kinds of variety on the effectiveness continuum uses the strengths of a company's current development process to arrive at the very best it is capable of *today*—but not as good as it will be capable of tomorrow. This is a subtle procedure, one that requires a series of trade-offs.

Designers and engineers have a powerful leadership role to play. They are in an ideal position to keep their eye on the big picture, instituting policies that can take a substantial chunk out of cost and complexity. They are also the only ones who can put such policies into day-to-day practice—exercising high- and low-level diplomacies to short circuit negative variety triggered by unwarranted parts variation. Strategic cost-cutting used to mean finding cheaper, better ways to manufacture. Now it also means finding cheaper, better ways to design and engineer products.

With this chapter, we complete the second section of this book, with its survey of VEP principles, and the problems they are designed to address. In the six chapters of the next section we present the VEP methodology, including details on how to prepare for a successful VEP launch and how to conduct the analysis at VEP's heart.

PART II
VEP
The Methodology

CHAPTER 6

The VEP Methodology Stage 1: Getting Ready To Launch

Companies with the lowest complexity grow 30% to 40% faster than their average competitors.

— *Bain & Company, 2014*

Resistance to change is a fact of corporate life. After decades of product expansion and building the skills to manage the ensuing complexity, managers rarely welcome the notion of disentangling the organizational complexity that masks huge parts inventories. But that same complexity also masks the opportunity to maximize customer selection. For most companies, an authentic commitment to variety effectiveness means—a lot of work.

It also means a lot of reward. For example, experts project that if car makers cut parts cost by just 3%, they would triple the industry's operating profits. For this to happen, you need a powerful methodology that systematically roots out the full range of negative causes, builds new positive policies and practices, and keeps the vision clear and compelling. *VEP: Variety Effectiveness Process is that method.*

In this chapter, we introduce the VEP Methodology—first discussing the matter of scope, then providing an overview of the four stages, and concluding with the details of the first stage.

Deciding On Scope

Embarking on the VEP journey requires your company to commit resources. The question is how much: What is the right scope for your organization? There are two options: the *discrete approach* or the *deep-dive approach*. Whichever approach you choose, VEP's goal remains the same:

lower costs and complexity dramatically, directing all product-linked expenditures toward *a single source of costs*—the part.

The Discrete Approach

The *discrete approach* is more of a tactic than a strategy—it is not comprehensive. It focuses on winning some battles against negative variety but not the war. In it, a task force of experts (mostly engineers and marketing representatives) is first trained in the VEP Methodology and then they select a focus—a single product or product line for analysis and reduction.

With only that product series in mind, for example, the task force largely focuses on parts reduction, with other factors entering the discussion as they surface. As you will learn in detail in Chapter Nine, parts reduction is one of the three goals of VEP's central reduction tool: *The 3-View Analysis.*

Depending on a number of factors, a single implementation cycle of this nature takes two to six months to complete. Afterwards, the team may elect to cycle back to another single-product focus for parts reductions—and then another and another, methodically lowering inventory levels, product by product.

Another option for the discrete approach to VEP is a focus on products with special characteristics. For example, VEP activity could target products at or near the end of their life cycle—products about to be retired or already discontinued. For discontinued products, eliminate all dedicated parts associated with it. For products close to their retirement, determine a reasonable end date and then schedule the elimination of any dedicated parts from inventory. Both scenarios can also trigger substantial reduction in storage space and any linked processes. Some companies will find this approach less disruptive than trying to address parts within products in active markets; it also encourages attention to the often neglected issue of obsolescence.

Some companies may prefer to address their VEP attentions under a discrete approach to their SKU portfolio and bring about reductions in it. Although far greater levels of simplicity and associated cost savings are available by attacking the parts level (deep-dive approach), beginning with SKUs may offer a softer and therefore more appealing entry point. The caution here is that removing SKUs does not automatically remove underlying causes of negative variety.

Whatever your focus, the discrete approach is a perfectly valid improvement tactic, especially in the face of limited company resources.

Targeted and sustained, this effort can—over time—result in considerable reductions in parts count on a product-by-product basis. In addition, it can measurably lower the number of processes and control points, especially if policy upgrades and revisions are included in the scope. This low-profile, contained tactic is also a good choice when a company wants to take a wait-and-see attitude before making an all-out VEP commitment.

The Deep-Dive Approach

The second option is an all-out attack on negative variety. We call this the *deep-dive approach*. It embraces variety effectiveness as a corporate vision and VEP as the company-wide strategy to reach it. This approach aims at building awareness across the entire organization and achieving dramatic reductions in parts inventory, production processes, and control points as well as installing a new set of practices that support positive variety by revising policy in all the core organizational functions—accounting, marketing and sales included.

Training is a key element in the deep-dive approach. People need to be readied. Since the deep-dive approach is a long-term effort that results in the eventual scrutiny of the company's full line of products, a full array of teams has to be activated and trained in the VEP Methodology (see below). These teams represent all departments. Immediate and measurable results can happen quickly, within the first 30 to 90 days, and more widespread changes continue to unfold over a period of 8 to 18 months.

We use the deep-dive approach in the balance of this book. In this chapter, we overview VEP's four stages and then study Stage 1 in detail. Parts Unlimited Inc. (PUI), our prototype company, continues as our case study.

Overview Of The VEP Method

The Variety Effectiveness Process is a systematic, team-based methodology directed at maintaining or expanding customer selection while reducing negative variety in products, parts, processes, and control points—and preventing their recurrence. Its goal is to lower costs dramatically and de-complicate internal systems while maximizing a company's ability to respond successfully to the demands of the market.

As a systematic approach, VEP spells out what needs to change, how, and in what order. The method is divided into four stages:

Stage 1: Plan and Prepare for an Effective Implementation
Organize and train your VEP Teams. Decide where to begin: Select your targeted series. Update and streamline your Parts Classification System.

Stage 2: Identify Valid Reduction Opportunities
Applying a powerful set of six VEP Analysis Tools (Six VATs), VEP Teams search out negative variety and develop valid proposals for reducing it.

Stage 3: Coordinate, Prioritize, and Schedule Reduction Opportunities
Compile, align, and schedule all valid reduction recommendations into an integrated plan of change.

Stage 4: Implement Improvements and Sustain a VEP Mindset
Implement valid improvement proposals, establish prevention procedures to avoid falling back into previous bad habits, and sustain.

This four-stage process is designed to help your company:

- Assess its true capacity for change and mobilize the needed resources.
- Implement a change protocol that systematically turns over the assorted rocks under which negative variety hides.
- Separate valid reduction proposals into ones with low resistance to change that can be implemented quickly and in the near term—and those that require a deeper stratum of support and planning to effect.
- Move through the implementation process so the designated changes are successful and payoffs are commensurate with required efforts.

From the first, you look for ways to simplify company systems and reduce parts inventories, even as you expand product offerings to customers. In parallel, there is an ongoing process to root out and prevent negative variety by codifying new policies and practices that strengthen positive variety and deter the recurrence of negative variety.

VEP's four-stage method provides you and your company with a road map for change, showing the organization where it needs to go and how to get there, reliably and safely. You can follow this idealized route until you get where you want to go—or modify and adapt the route by factoring in the special strengths and idiosyncrasies of your enterprise in order to fashion useful shortcuts. See Figure 6.1 for an overview of the four stages of the VEP Methodology (Stage 1, under discussion in this chapter, is highlighted in gray).

Figure 6.1. **VEP Methodology: Stage 1**

STAGE 1 Plan and Prepare for an Effective VEP Implementation	STAGE 2 Identify Reduction Opportunities by Applying the Six VATs	STAGE 3 Coordinate and Schedule Reduction Proposals	STAGE 4 Implement Improvements and Sustain a VEP Mindset
Step 1 Select a Steering Team that then sets up the other VEP Teams *(Management)*	**3-View Analysis Teams** { **Step 1a** Undertake a VEP analysis of market offerings and their characteristics and make reduction proposals *(Market Analysis Team)* **Step 1b** Undertake a VEP analysis of parts as part of the product architecture and make reduction proposals *(Product Structure Analysis Team)* **Step 1c** Undertake a VEP analysis of parts by parts type and make reduction proposals *(Parts Type Analysis Team)*	**Step 1** Coordinate and consolidate reduction proposals *(Steering Team)*	**Step 1** Implement approved reduction proposals *(all VEP Teams)*
Step 2 Conduct VEP training for teams and begin general awareness training *(Education and Methods Team)*	**Step 2*** Undertake a VEP analysis of transactions that support parts, products, and market offerings and make reduction proposals *(Control Points Reduction Team)*	**Step 2** Qualify, approve, and prioritize reduction proposals *(Steering Team)*	**Step 2** Set up a VEP Preventative Monitoring Calendar and continue to educate a VEP mindset *(all VEP Teams)*
Step 3a Find and reduce parts with low-resistance to change *(Early Victories Team)* **Step 3b** Assess, clean up, and upgrade parts classification system *(Parts Type Analysis Team)* **Step 3c** Begin to analyze and revise company policies and practices *(Policy Analysis Team)*	**Step 3**** Undertake a VEP analysis of processes and make reduction proposals *(Process Reduction Team)*	**Step 3** Schedule approved reduction proposals on a VEP Implementation Calendar *(Steering Team)*	
Step 4 Target a priority product as the starting point for Stage 2 reduction analysis *(Steering Team)*			

* The Control Points Reduction Team works independent from other VEP Teams—and can start its work in Stage 1 (see Chapter Ten).
** This team is an optional step, dependent on current levels of operational complexity and improvement.

Details Of Stage 1:
Prepare For An Effective Implementation

The first VEP stage begins when senior management publicly commits company resources to a strategy of variety effectiveness. At PUI, this happened when CEO Tom Vargas and his board made the decision to pursue VEP and sent a memo to all employees announcing that launch (Figure 6.2).

Figure 6.2. **Memo: Company Commitment to VEP**

To: All PUI Employees
From: T. M. Vargas, CEO

RE: Kick-Off of the Variety Effectiveness Process at PUI

As we discussed last month at the company meeting, our parts inventory costs are exploding and must be brought under control.

Just before the December holidays, the entire management team attended a one-day briefing on a powerful, systematic methodology aimed at reducing high levels of parts inventory, de-complicating the organization, and providing our customers with a better product selection at a more competitive price. We anticipate a significant growth in our profit margins as a result. The method is called: VEP/Variety Effectiveness Process®.

After further research and discussion involving many of you, we are ready to commit. *On February 22, we kick off VEP at PUI.*

Our goal is a challenging one: By December 31, we will reduce our parts inventory by 40%, production processes by 30%, and control points by 25%

Here is the calendar for the next two months:
        February 2   : VEP Steering Team and other VEP Teams are formed.
        February 15 : All VEP Teams attend 3-day VEP Training Course.
        February 28 : VEP Kick-Off. Teams begin their work.
        March 15   : First VEP Awareness Session—all employees.

Our company is committing substantial time and money to get what is called negative variety under control and our organization de-complicated. I need you to get involved and help make it happen. PUI needs your knowledge, skill, and enthusiasm. It won't be easy—but it will be worth it.

Working together, our product designs will improve, our internal systems will become streamlined, our customers will be happier—and our profits will improve.

Thank you for your support in this very important initiative.

Once the company has announced its commitment, it undertakes the three main tasks of Stage 1: select VEP Teams; train VEP Teams and the general workforce; and select a priority product series for VEP analysis.

VEP Teams

A comprehensive VEP implementation is carried out through a set of cross-functional teams which we categorize as leadership, analysis, and promotion and support. Figures 6.3 lists the VEP Teams in the approximate order of their formation.

Figure 6.3. **VEP Teams for a Deep-Dive Implementation**

VEP TEAMS	OBJECTIVES
1. Steering Team	• Set up other VEP Teams. • Provide leadership, direction, and support to the overall project. • Review, coordinate, and finalize reduction recommendations from VEP Teams. • Serve as liaison to Senior Management.
2. Early Victories Team	• Identify parts and processes with low or no resistance to change and reduce these, respectively, by a targeted amount in the first 90 days.
3. Education and Methods Team	• Arrange for VEP Team Training. • Conduct VEP Awareness Sessions. • Assist teams in using and adapting the VEP Method. • Promote involvement and publicize successes.
4. Policy Team	• Identify/review/revise company policies and practices (formal and informal) so that they support positive variety and avoid negative variety in the future.
5. Parts Type Analysis Team *(The 3-View Analysis Teams)*	• Assess existing parts classification system; validate or upgrade as needed. Then develop and implement attribute templates and input new data accordingly. • Identify opportunities for reduction in parts commodities across the company's entire parts universe.
6. Market Analysis Team *(The 3-View Analysis Teams)*	• Identify reduction opportunities in company's existing product offerings that do not endanger the company's market position or limit customer selection.
7. Product Structure Analysis Team *(The 3-View Analysis Teams)*	• Identify reduction opportunities on model-specific parts and structures across all product lines, based on the analysis of product architecture.
8. Control Points Reduction Team	• Identify and reduce all transactions (paper, electronic or otherwise) that supports the design, procurement, sorting, retrieval, scheduling, inspection of parts, etc.
9. Process Reduction Team (optional)	• Identify and reduce unnecessary processes in operations. • Note: Many unnecessary and redundant processes will have already been eliminated by the parallel reductions in parts, product or market offering of the 3-VEP Analysis Teams. In some companies, that means a production process reduction team will not be required.

VEP Leadership Teams

Two teams serve in a leadership capacity for the VEP process: the Steering Team and the Policy Team.

Steering Team. Multi-team implementations need clear direction and ongoing integration. The first team formed is the *VEP Steering Team*. Its purpose is to provide leadership, direction, backing, and alignment to the overall process. Later, the Steering Team will serve as a clearinghouse for the improvement recommendations submitted by the various analysis teams—regularly reviewing, coordinating, and, when the time comes, prioritizing reduction proposals. This team also serves as gatekeeper, regulating the rate at which change/improvement is introduced into company systems. This is one of its most important functions.

Steering Team members are chosen by senior management not only to sit on the Steering Team—but also to form and then lead one of the other VEP Teams. The Steering Team itself is usually led by an engineer who reports directly to senior managers and keeps them and other key stakeholders informed and on board.

Policy Team. Once a company understands that *parts trigger cost*, the battle lines are drawn. While the VEP analysis teams (described in detail below) are charged with scrutinizing the causes of the company's product and parts explosion, the VEP Policy Team goes deeper. Its mission is to identify and name those official and unofficial policies and practices that, however inadvertently, promote negative variety—to root out causes buried deep in the status quo.

Discussed at length in Chapter Four, this investigation requires digging into the fabric of how the company conducts its business. Because of the power of policy, no department can be spared. Change on the policy level, therefore, is challenging. For one thing, policy changes can shake the status quo to its foundation. For another, though bad policy may be inadvertent, it nonetheless casts a long shadow. A well-working Policy Team fuses discerning examination with substantial political sensitivity. It is advisable, therefore, that the team leader be a senior executive—maybe even the VEP champion—a person in a position to lubricate the inquiry and facilitate the needed changes that advance policy improvement.

The work of policy analysis begins early in Stage 1, on the heels of the VEP Team Training. The mechanics are straightforward. Team members collect and scrutinize the company's formal and informal policies and tag the ones that appear to provoke negative variety. The grid of five organizational areas (Accounting, Marketing and Sales, Product Development, IT/Data Systems, and Operations) is the guide, along with the seventeen triggers found in Chapter Four (see Figure 4.2). Using these triggers as models, this team also pinpoints policies that support positive variety.

Once the policy roots of a problem are found, the team recommends specific revisions and upgrades, with each member bringing the pertinent ones to the attention of his/her respective department for review and input. If a policy is particularly troublesome, the department as a whole engages in an open-ended dialogue on what to change; those recommendations are sent back to the Policy Team.

Expect some cultural push back, since policy is where the truth of the company resides. The Policy Team, under the auspices of the senior manager who leads them, must undertake the delicate process of shepherding improved policies through the system so that stakeholders are kept on board. In this sense, the team plays a key role in forging the company's future direction toward effective variety. By design or not, policy is the driver. Intentions may be pure but people can still experience the efforts of this team as pushy and not satisfying. If you sense this, rely more on the open-ended dialogue option.

Once the work of this team takes hold and starts rolling, the process becomes iterative and all the policy bases get covered. Along the way, senior management officially sanctions new policies. As a stabilizing step, once a crop of improved policies is in place, the Policy Team designates a small group in each department to watchdog for compliance and incursions.

VEP Analysis Teams

The heart of the VEP method is an analysis process that results in specific improvement proposals for optimizing products and minimizing parts inventory levels. This process, in the main, is the shared effort of three separate but linked teams, each named for the particular analysis window it opens to scrutinize the company's offerings:

1. Market Analysis Team
2. Product Structure Analysis Team
3. Parts Type Analysis Team

Collectively referred to as the *3-View Analysis*, these three teams are the engine that drives the VEP approach. Between them, products and parts are analyzed and specific opportunities for reduction are developed.

We briefly define them below; in Chapter Nine we delve deeply into each. Two further teams are optional: the Control Points Reduction Team and the Process Reduction Team. They are discussed at the end of this section.

1. Market Analysis Team

The purpose of the *Market Analysis Team* is to analyze company products from a marketing perspective and propose reductions that eliminate unneeded variety—without limiting customer selection or endangering the company's market position.

In essence, this requires cataloging and comparing company offerings in order to identify instances of overlap, redundancy, indistinctness, and other anomalies. This analysis can extend across the full scope of product features and functionality and into distinctions in nomenclature and distribution channels. Team members, therefore, must have sufficient experience to handle a wide range of questions and possibilities. Minimally, the team should include representation from Marketing (also a good choice for team leader), Design Engineering, Manufacturing Engineering, and Finance.

2. Product Structure Analysis Team

The *Product Structure Analysis Team* studies the architecture of company product offerings for reduction possibilities. This team examines how products are put together and how their discrete functions are realized. Team members, for example, could conclude that a complex housing and bracket assembly can be simplified and five part numbers eliminated by combining two of the five elements in that assembly.

Due to this team's heavy engineering demands, its core composition should include design engineers as well as representatives from Manufacturing Engineering, Drafting, Purchasing, and Marketing.

3. Parts Type Analysis Team

The *Parts Type Analysis Team,* as its name suggests, concentrates on negative variety found in parts (also known as parts types), as exemplified in housings, fasteners, windshield wipers, knobs, labels, lids, brackets, side mirrors, handles, washers, steering wheels, gaskets, side panels, hinges, facings, and so on.

As with the other analytical teams, members work hard to develop reduction proposals that do not compromise customer selection. In a comprehensive VEP implementation, this team eventually surveys all the parts types in the company's parts universe. Its first task, however, is to assess and upgrade the organization's parts classification system and develop attribute templates (discussed at length in the next chapter). For that reason, this team is formed at the outset of the VEP process.

As with the Product Structure Analysis Team, this team's strong engineering focus is the reason it includes representatives from Design and Manufacturing Engineering; typically, people from Systems, Purchasing, and Marketing also serve.

As mentioned, two other VEP analyses may also be undertaken: control points and processes, overviewed here and discussed in detail in Chapter Ten.

- **Control Points Reduction Team (optional)**

A *control point* is any transaction—paper or electronic—that supports a product, a part or a process. This includes drawings, purchase orders, as well as transactions associated with bills of material—receiving, retrieving, and inspecting parts; billing; customer service; and all manner of memos. The mandate of this team is to identify these and reduce their number.

Because of the extensive number of control points in most companies, the Control Points Reduction Team often includes representatives from every department: Purchasing, Systems, Drafting, Design Engineering, Materials, Manufacturing Engineering, Finance, Quality, Marketing, Human Resources, Customer Service, etc. Any team member with good knowledge of the company and with strong leadership skills is suitable to lead this team.

- **Process Reduction Team (optional)**

The second optional team is the Process Reduction Team, which seeks to reduce the number of required processes as a means of simplifying operations. If your business is manufacturing-based, the larger part of these will be production processes. If your business is service-based, these will be transactional in nature.

Whichever the case, this team is optional for two reasons. First, while process reduction is critical to de-complicating the organization, this reduction can be (and often is) done as part of the work of the Product Structure and Parts Type Analysis Teams. That is, as products are simplified and parts reduced, associated production processes are tagged for elimination as a matter of course. No special extra effort is needed.

The second reason a separate process reduction team may not be needed is because your company has already done substantial work in process reduction, as PUI did with cell design and in dies and fixture standardization prior to its VEP implementation. But if your company is new to process improvement, it may well benefit from a separate effort—and therefore a separate team—targeting those reductions. You decide.

As before, the team starts by standardizing the nomenclature for processes; soon they are ready to search for reduction opportunities. The major groups represented on this team typically include: Manufacturing Engineering, Marketing, Operations, Material Handling, Machine Shop, and Stores.

VEP Support Teams

The changes brought about by VEP are exceedingly positive and long-lasting—but don't expect everyone to be enthusiastic from the outset. As in any profound organizational change, there will be no shortage of naysayers, skeptics, and resisters. While some negativity is to be expected, it can nonetheless be destructive during the start-up. Steps taken early on can harness or even diffuse the pessimism. This purpose is shared by the two teams responsible for promotion and support: the Early Victories Team and the Education and Methods Team.

Early Victories Team. The *Early Victories Team* moves into action quickly and has a short life span of three months—90 days. Its target is very specific: a) Identify parts and processes with low or no resistance to change, and b) reduce their number by a pre-set amount within the first 90 days of the implementation. A specific target, for example, might be: *Reduce parts numbers with low or no resistance to change by 1,000 and production processes by 100 in the next 90 days—by September 15th.*

The idea here is for this team to score the easy successes, early in the VEP process—and then, with the assistance of the Education and Methods Team (below), get them publicized. Early victories keep the skeptics at bay while the harder work of preparing for and beginning in-depth analysis gets under way. This is an end-run around those who may want to torpedo the project in its infancy.

For 90 days, this team works on getting rid of the easy stuff. Cleaning up the parts database of obsolete parts numbers is an obvious and necessary part of this. As these parts numbers are eliminated, certain corresponding production processes are eliminated as well. This team usually includes full-company representation: Purchasing, Marketing, Sales, Drafting, Design and Manufacturing Engineering, IT Systems, Finance, Customer Service, Materials, Quality, Human Resources, Shipping and Receiving, etc. One immediate benefit of such a widely representative group is the likelihood that team members will tell their colleagues what VEP can do and why it is important. This simple word-of-mouth promotion goes a long way in keeping the atmosphere around the fledgling implementation open and upbeat.

In one company, which eliminated more than a thousand parts numbers in the first three months, the Early Victories Team promoted its efforts through an imaginative device called "The Chuck Wagon"—so named because team members would "chuck" eliminated parts onto a cart in the company lobby. (See Figure 6.4 for illustration. Notice the plastic arm at the top of the wagon to "catch" unnecessary parts.)

Figure 6.4. **The Chuck Wagon: Early Victories Team**

Education and Methods Team. In VEP, the majority of the work force does not serve on a team; yet their cooperation is vitally important. That cooperation depends on their knowing what is expected of them. This is the work of the *Education and Methods Team,* usually comprised of members of the training function (trainers, facilitators, coaches, and internal consultants) and Human Resources.

This team's mandate is to make sure everyone knows what VEP is, why it is important, what the broad-stroke timeline is, and how everyone can contribute to its success. This is done in part through the series of *VEP Awareness Sessions* this team conducts across the company in the opening months of the implementation. In addition to these sessions, the Education and Methods Team schedules and conducts in-depth training for all VEP teams. (See the Resource Section in the Appendix for VEP Training Materials and Services.) This usually requires a multi-day workshop so team members learn basic VEP principles and concepts, the big picture, and the particular mission of each individual team; teams learn and practice the technical procedures and supporting team behaviors. As a result, they gain a strong sense of the purpose and direction of the implementation, and build confidence in their respective team tasks and in the VEP process in general.

The Education and Methods Team is also responsible for promoting VEP results and highlighting the positive benefits. VEP bulletin boards, for example, spotlight the work of individual teams as well as collective results. A column in the company newsletter can also be of great help in keeping everyone on board and alert. The Education and Methods Team looks for and develops such opportunities.

A third function of this very important team is to help the Steering Team customize VEP methods to fit the needs of the company more exactly, working with all other VEP teams to clarify targets and objectives and keep on track. The Education and Methods Team may, for example, work with the Control Points Reduction Team in adapting the VEP Parts Index so it can capture the number and frequency of the forms used in Purchasing and Finance. Or it may help the Steering Team think through its timing issues. In short, the Education and Methods Team is the central knowledge and information base of the project, providing support to all the players as the implementation goes forward.

When it comes time to organize your teams, adjust the details in these descriptions to better suit your company—but make sure to keep the core purpose clear. You will also have other questions about, for example, the time commitment members of each VEP Team are expected to make. Figure 6.5 shows you the way one organization addressed those issues and more in the detailed *VEP Team Fact Sheet* it developed.

Figure 6.5. **Sample: VEP Team Configuration, with Time Commitments**

VEP TEAMS	GOALS	TEAM MEMBERS	TIME COMMITMENT
1. Steering Team	• Oversee and monitor the activities of all VEP Teams and maintain their focus and continuity relative to overall project. • Revise timelines as needed and identify/request supporting resources as required. • Evaluate and prioritize each team's reduction recommendations, coordinate them into a final plan and oversee their deployment. **Output:** A final implementation plan, developed in conjunction with senior management, of improvement proposals from the VEP Teams.	*Design Engineering/LEAD* • Marketing* • Purchasing* • Accounting* • IT/Data Systems* • Education and Training* * *Each member of the Steering Team also serves as the LEAD on one of the other VEP Teams.*	**Overall Time Commitment:** 14 months **Weekly Time Commitment:** • Team Lead: 8%-10% • Team Member: 8%-10% • Admin Assistant: 5%-8%
2. Early Victories Team	• Visibly demonstrate to employees and other VEP Teams that parts count can be reduced without negatively impacting customer selection. • Provide a morale boost to all employees through tangible evidence of early VEP successes. **Output:** The actual reduction of parts count by 1000 within the first 90 days by attacking parts with low or no resistance to change.	*Purchasing/LEAD* • IT/Data Systems • Marketing • Engineering • Materials • Finance • Drafting	**Overall Time Commitment:** 3 months **Weekly Time Commitment:** • Team Lead: 8%-10% • Team Member: 8%-10%

VEP TEAMS	GOALS	TEAM MEMBERS	TIME COMMITMENT
3. Education and Methods Team	• Train the VEP Teams in the VEP Methodology, incorporating special steps, aligned with the company culture and other special needs of the site. • Conduct twice weekly VEP Awareness Session for the general workforce. • Provide re-enforcement training and support to individual teams as required. • Promote and publicize VEP successes. **Output**: The ongoing training and support of VEP Teams.	*Education and Training Manager/ LEAD* • Education & Training Staff	**Overall Time Commitment:** 4 months (then reduced but ongoing) **Weekly Time Commitment:** • Team Lead: 10%-15% • Team Member: 10%-15%
4. Market Analysis Team	• Identify opportunities for reductions in the company's product universe from a market perspective by applying the 6-VATs. **Output**: Produce a prioritized list of reduction proposals in our existing product line while protecting the company's market position and customer selection.	*Marketing/LEAD* • Design Engineering • Mfg. Engineering • Finance	**Overall Time Commitment:** 5 to 8 months **Weekly Time Commitment:** • Team Lead: 10%-15% • Team Member: 8%-12%
5. Product Structure Team	• Identify opportunities for parts reduction on model-specific parts across all product lines by developing parts indices and applying visible layout analysis and the 6-VATs. **Output**: A prioritized list of reduction proposals for our product lines while protecting the company's market position and customer selection.	*Design Engineering/LEAD* • Purchasing • Mfg. Engineering • Drafting	**Overall Time Commitment:** 10-12 months **Weekly Time Commitment:** • Team Lead: 10%-15% • Team Member: 8%-12%

VEP TEAMS	GOALS	TEAM MEMBERS	TIME COMMITMENT
6. Parts Type Team	• Identify opportunities for reduction in parts types across the company's parts universe by determining attribute overlap and redundancy and by applying the 6-VATs. **Output**: A prioritized list of proposals that reduce the number of parts in our parts universe while protecting the company's market position and customer selection.	*Design Engineering/LEAD* • Systems • Mfg. Engineering • Marketing	**Overall Time Commitment:** 10-12 months **Weekly Time Commitment:** • Team Lead: 10%-15% • Team Member: 8%-12%
7. Control Points Team	• Identify opportunities for reducing transactions (and linked paperwork) associated with the design, acquisition, work upon, inspection, storage, retrieval, handling, and otherwise management of parts. **Output:** A Master Control Points Index and regular monitoring of control points proliferation.	*Accounting/LEAD* • IT/Data Systems Drafting • Engineering Operations • Marketing	**Overall Time Commitment:** 5 months **Weekly Time Commitment:** • Team Lead: 10%-15% • Team Member: 5%-10%
8. Policy Team	• Identify and review existing company policies and practices (formal and informal) in order to determine the changes needed to curb or avoid future negative variety. **Output**: Improved/revised policies that are accepted and implemented buy associated departments.	*CEO/LEAD* • The Steering Team	**Overall Time Commitment:** Spans entire VEP Roll Out **Weekly Time Commitment:** • Team Lead: 10%-12% • Team Member: 8%-12% • Admin Assistant: 5%-8%

Select Your Starting Point

Stage 1 of the VEP Method includes one more vital preparation step: Determine where to begin your VEP analysis. This is a pivotal decision, one that answers the following questions:

- Should the 3-View Analysis start with the same product or different ones?
- If it is the same product, should the focus be on a single product or the entire product line?
- Would it be better to start analyzing several associated product lines simultaneously and set a strong pace—or stay with a single product line?
- Whichever the case, how can the teams be confident that their efforts—whether joint or separate—will pay off?

In the United States, our instant answer to everything is: Just Do It! In the best *ready, fire, aim* tradition of American rugged individualism, preparation and planning are needless obstructions. In VEP, nothing could be further from the truth. Instead, VEP offers a systematic way to plan and prepare for a successful launch—a process for selecting the single product series most likely to benefit the company through VEP analysis. That single series is called the *targeted line*. This is the starting point.

Even if you think you already know your product line choice, we recommend that you engage in the following qualifying process to validate your choice. Yes, it is that important.

The Qualifying Procedure

The VEP Steering Team is responsible for selecting the targeted line. To do so, members ask and answer this question: *Of all our product lines, which one will give us the most bang—the biggest return—for our analysis buck?* Here is an overview of that process and how it unfolds.

- Determine a set of assessment criteria and a weighting formula
- Apply this scoring formula to each of your product lines
- Rank order your product lines based on resulting individual scores

The product line with the highest score wins: It becomes the *targeted line* and the starting point for VEP analysis.

Determining Your Targeted Product Line. What are the best assessment criteria for determining the targeted series—criteria that will produce the biggest payoff for our VEP efforts?

Ask three departments, get three different answers. Engineers, for example, want to use the level of structure complexity and/or ratio of dedicated-to-shared parts as expressed in the VEP Parts Index. "That way," they say, "we'll improve the products that give us the biggest headaches." Marketing prefers to apply the level of revenue contribution as a key criterion. Improving those products, that logic goes, will make the company's return-on-sales jump even higher. Operations always goes with quantity of shipped units and/or number of required dies and fixtures, in keeping with its goal to make day-to-day life on the shop floor a little easier. And don't forget about the position of a product on the life-cycle continuum, another important gauge. And there are more.

The point is: All these factors are worth considering. It is the task of your Steering Team to do just that: Identify as many of them as possible and then select the ones most relevant to your organization; this positions for the next step—qualifying likely candidates. A word of caution, related to the true cost discussion in Chapter Three: Avoid criteria linked to the traditional accounting formula factors of Labor + Material + Overhead (L+M+O). They tell us nothing about the complexity of a product series and will only muddy the waters.

The Steering Team is wise to begin with simple brainstorming. Using a flip chart, team members brainstorm a list of possible criteria (no evaluation, please). Each person then ranks the items on that list according to their own preferences: which criterion is most important, which is second, then third, and so on. Next, people discuss and justify their individual preferences and then, through consensus, arrive at a joint, pared-down list.

Or you can do it in linked circles, using a modified relations diagram (Figure 6.6). Either way, you are answering the central question: "Which qualifying criteria will give us the biggest bang for our VEP analysis buck?" The relations diagram format gives you the advantage of developing first- and second-level responses (primary and secondary criteria)—visibly for all to see and calibrate. If you do this with sticky notes, you can move those criteria around, at will, until a configuration that is right for your company emerges.

Figure 6.6. **Modified Relations Diagram**

With the criteria selected, clarify the categories and segment them into subsets as needed. Next, develop a numerical scoring scale and weighting value (see below) to apply to each product series. In this way, you can mathematically compare one candidate with another. When you complete this, all your company's offerings will be positioned on a summative scale (rank-order continuum)—with the most qualified product series at the top. That top series becomes the priority and starting point for the VEP analysis. It becomes your targeted series.

Designating your targeted series as the starting point is the link to the next round of VEP activity—the analysis itself. Anchored to this single product line, each of the 3-View Analysis Teams begins its respective work, using the same targeted series as the jumping-off point. This keeps the focus of the VEP implementation sharp and clear, preventing teams from slipping into the murkiness of too many choices, a symptom of the very complexity they are attempting to unravel.

How PUI Did It. Here is how the PUI Steering Team accomplished the selection. First, team members used the above process to select four qualifying criteria for each product series: 1) Expected Growth; 2) Sales Revenue; 3) Dedicated Parts Per Product Series; and 4) Ratio of Dedicated Parts To Total Parts.

Using the simple scoring scale below, the team then weighted the importance of those four categories across the company's product universe.

In this way, they identified their top five product series candidates: product lines 7, 53, 73, 11, and 148. In Figure 6.7 you can see the partial results of PUI's qualifying process. The targeted product line is Series 7.

Figure 6.7. PUI's Top Five Series: Score Sheet for Pre-Set Qualifying Criteria

SERIES	EXPECTED GROWTH	SALES REVENUE	DEDICATED PARTS INDEX	UNIQUE TO TOTAL PARTS RATIO	TOTAL WEIGHTING	NOTES		
						HIGH 8-9-10	MEDIUM 4-5-6-7	LOW 1-2-3
TARGETED SERIES 7	10	8	10	8	36	Continued growth expected. High dedicated parts. High shipped units.		
53	3	9	8	10	30	Moderate growth expected. High shipped units and parts.		
73	6	7	7	6	26	Moderate growth expected. High dedicated parts. Reasonably high shipped units.		
11	2	9	5	8	25	Tremendous revenue growth expected in three years. Shipped units will rise rapidly beginning next year. Fairly high dedicated/unique parts.		
148	2	5	5	10	22	Revenue to grow very slowly. Shipped units fairly high. Total parts fairly high. Important series to process industry.		

Why Qualify? We strongly recommend that a company ready to embark upon the VEP process go through this qualifying process, even if you think a priority product is obvious—and, even if you decide to pursue the discrete approach instead of the deep-dive. Either way, you benefit from systematically selecting your focus.

There are several reasons for this. First, even if you "know" which of your product series is most qualified, going through the qualifying process helps you document your thinking and the logic by which you made your choice. Second, in doing so, you revisit your key issues and may find you need to adjust your preset choice, one way or another. Only a systematic approach allows you to do this. Finally, if you need to go to management for approval or added resources, VEP's qualifying process helps you substantiate and validate your thinking.

With this selection made, you are ready to get to work.

Next In VEP's Stage 1

Stage 1, however, is not yet complete. In order to ensure that the VEP analysis work of Stage 2 produces the greatest return for your investment of time, energy, and commitment, the Parts Type Team must first undertake the arduous work of making your classification systems VEP-capable.

CHAPTER 7

Creating A VEP-Capable Classification System

The impact of reducing the number of parts—at the part level—from the overall manufacturing system shaves years off your improvement journey.

— ERIC LAIL, Transportation Insight

In Chapter Two, you learned that the average cost of producing a single new part is rarely less than $4,000—and can be as high as $40,000 or more. Another study of new parts introduction across 500 companies showed that the odds are 1 in 11 that an existing part is similar enough to be used instead of a new part. Whether an existing part is used—or a new one is created—rests squarely on the company's ability to access and use its classification system. Can you quickly and efficiently validate—or refute—the need for a part by accessing existing data?

Here is a case in point, a true one and not for the faint of heart. If there are small children present, we suggest you cover their ears.

Venerable Chair Company

Eric Lail, now Vice President of Client Services and Continuous Improvement at Transportation Insight, a third-party logistics consulting firm, has long been a supporter of *Smart Simple Design* and the VEP mindset. In this story, Eric was working for Venerable Chair Company (VCC), an alias for an actual, highly regarded furniture manufacturer.

Established in 1911 and still operating out of its original facility, VCC was reaping the whirlwind of almost two centuries of ever-expanding customer choice. To call their inventory excessive is an understatement: their five-level, 600,000 square foot facility was stuffed, for starters, with 865,000 wood parts—and these were just the ones in wood. A typical Y-type business, VCC was a respected maker of high-end goods but with profit margins relentlessly shrinking from the glut of proliferating products and parts.

Eric came on board as general manager, responsible for manufacturing and product engineering. It did not take him long to realize that he had inherited a legacy of chronic complexity. Estimated lead time for any upholstered product was 12 weeks—but actual on-time delivery hovered around 60%. VCC prided itself on customer happiness, offering 100 standard paint finishes and unlimited custom paint options—even though custom paint triggered an almost non-existent 1% of sales. Upon his further investigation, Eric discovered that the supervisor of the paint finish area spent 50% of his time on custom paint. Yes, you read it right: the profit-producing 99% of paint got only 50% of his attention. Even then, 90% of total sales stemmed from only ten of the 100 standard colors. At VCC, the common 80/20 rule was more like 90/10 and fertile ground for negative variety.

Juanita Hicks to the Rescue

But paint was only one aspect of the problem. How, we must ask, did a company with this magnitude of parts number proliferation handle the pulling of materials? At VCC, the answer was 30-year employee Juanita Hicks—and she supervised a group of 26 employees (out of full work force of 500) whose sole job was pulling parts. There was no computer data to help them, no 5S solutions, no visual devices—just Juanita's prodigious memory.

For the next two years, stabilizing the parts storage challenge was a critical improvement initiative mandated by Eric at VCC. Inroads were made, but in the absence of a committed and educated implementation of VEP principles, progress was limited. Then catastrophe loomed. Mrs. Hicks was about to retire and take the bulk of her vital knowledge of parts location with her. Eric caught wind and, with his understanding of improvement and VEP, rushed to offer Juanita a deal. If she would agree to stay, all she would do was work with designers to pull *existing* parts for new models. During the ensuing two-year period, engineering had to have Juanita's approval before creating a new part—and new parts were, as a result, rarely needed.

Eric left a few years later, taking his VEP knowledge with him. VCC went on to be very successful for the next decade, leading the industry in profits, design, and lean initiatives. Unfortunately, for reasons other than those cited here, the company went into bankruptcy at the writing of this book.

The VCC story is sobering and has a familiar point: A process is only as good as the data on which it is based. Similarly, good analysis and good outcomes depend on good information and a good data system. This becomes even more compelling when a company faces parts and product proliferation at even a fraction of the level VCC grappled with. Without a data system that is accurate, relevant, complete, and accessible, the glut of complexity endangers any enterprise. The purpose of this chapter is to walk you through the process of assessing your own system and upgrading it as needed so that it is *VEP-capable*.

Taking On The Job

The prospect of cleaning up the parts classification system in your company may loom large. While this work cannot be undertaken lightly, it becomes much more manageable when it is:

- Started early in the VEP process—that is, in Stage 1
- Performed by technically qualified staff
- Tackled in workable increments

Make no mistake: An information-capable parts data system is necessary and essential, even if you were not choosing to pursue VEP. It is a necessary and essential good business practice. But if you are committed to uncovering and rooting out negative variety in your company, you cannot proceed without it. You must have the capability to access accurate, complete, and meaningful data. Much as with our prototype company, PUI, the data systems in many companies are—quite simply—a mess. Unaddressed, they get worse with each passing day. Companies pay an awful price for allowing this level of inaccuracy and confusion to continue. For some companies, this process may mean only a slight modification to their existing approach. For others, completing these three steps may seem roughly equivalent to hopping up Mt. McKinley on one foot.

All VEP Teams require sound data to do their work—none more than the three Analysis Teams we study in Chapter Nine, representing the heart of VEP's attack on negative variety. Each of these teams face a multitude of data-dependent decisions related to parts types, parts specifications, BOMs, product structure, quantity-on-hand, sourcing—as well as marketing facts and figures. The quality and scope of the company's parts classification system are crucial.

Let's look at trouble itself: the classification system at PUI.

A Case In Point: A Bulging, Bungling, Blundering Data System

PUI's parts and product classification system dates back to the company's founding in 1932. Though logical at the start, the system has grown inconsistent over time and is now rarely followed when new products are added. The introduction of a computerized system in 1970 tightened the classification framework but did not restore an underlying logic.

Since then, the now computerized system continued to inflate each and every time someone needed—or thought they needed—a new classification category, whether for a part, product or production process. The last time PUI updated its computer system, software, and classification approach was twenty years before the VEP launch.

During that time, anomalies, slight errors, and outright mistakes continued to accumulate in PUI's data systems. These went unnoticed, primarily because no one was looking. For two decades, no one noticed.

The arrival of VEP changed all that, offering PUI a straightforward process for dismantling and decomplicating its classification approach. That is not to say it was easy. But once PUI got educated to the staggering cost of an incapable system, the men and women in PUI's IT department—in alignment with engineering—dug in and undertook the task with gusto and success.

First, they noted the following broad areas of deficiency in their existing classification system:

- The terminology used at PUI to describe parts and products is inconsistent and confusing.
- The classification database is filled with data that are conflicting, duplicate, obsolete, inaccurate, and/or incomplete.
- Procedures for adding a new classification for a product or part are informal and inconsistent.
- Part numbers do not provide adequate attribute information for decision making.
- Product life cycle, sales, and other marketing-related data are not easily available to analysis teams.
- Class codes overlap and no rules are in place for assigning class codes to a new part or product.
- Rules that are in place are weakened by myriad exceptions.
- Information encoded in the 30-character alpha-numeric description

format is often inconsistent and is insufficient for making VEP decisions.
- PUI's data system has rarely, if ever, been maintained—no stated policy to do so exists.

When tackling the classification upgrade, PUI organized these into the following five key trouble areas.

Trouble Area 1: Part Number Prefixes

At the parts level, similar parts share the same prefix designation. For example, all springs carry <6318> as the first four digits in their designation. Digits after the first four are simply assigned sequentially as new springs are added to the parts universe. Unfortunately, this means that Engineering does not know—without sifting through endless drawing files—whether *Spring 6318-01* has any attribute in common with *Spring 6318-903*.

Trouble Area 2: Computer Codes

Twenty years ago, when PUI last updated its data system, it purchased state-of-the-art hardware and the latest ERP-based software. ERP required PUI to add more coding to the existing parts and product classification system which already relied on a great number of computer codes to support sales, production, procurement, and inventory control processes. The logic used to establish and maintain the new codes left a great deal to be desired. As a result, they did not add to the capability of PUI's classification system.

Trouble Area 3: Class Codes

PUI parts and products also carry computerized class codes that allow for the grouping of like items, such as sensor assemblies or enclosures. Unfortunately, many of these class code categories overlap since there are no concrete rules for assigning a new part or product to a particular class code category. Also:

- As of last year, the number of active class codes at PUI was 104. Figure 7.1 shows the codes as they existed *before* VEP, along with their descriptions.
- Codes 1 to 10 are reserved for completed product configurations.
- The bulk of the remaining codes are used for grouping parts, with a handful of codes for tools or fixtures.
- Drafting is responsible for assigning class codes to products and parts.

Figure 7.1. **Before VEP: PUI Active Class Codes (104 Total Codes)**

	CLASS CODE DESCRIPTION		CLASS CODE DESCRIPTION
1.*	Mechanical Stock	31.	Dialplates
2.	Non-Mechanical Stock	32.	Diaphragms, Gaskets & Seals
3.	Electronic Stock	33.	Miscellaneous Commercial Parts
4.	Electronic Non-Stock	34.	Insulators
5.	Replacement Sensor Stock	35.	Screw Machine Parts
6.	Replacement Sensor Non-Stock	36.	Wells & Connectors
7.	Electronic Parts & Accessories	37.	Metal Stampings
8.	Mechanical Parts & Accessories	38.	Switches, Parts For
9.	Process Stock	39.	Miscellaneous Metal Tubing
10.	Process Non-Stock	40.	Commercial Wire
11.	Miscellaneous Parts	41.	Packaging Materials
12.	Common Parts-Heads	42.	Capacitors
13.	Common Parts-Sensors	43.	Series 4 Head Assemblies
14.	Pressure Assembly	44.	Printed Circuit Boards
15.	Vacuum Assembly	45.	Relays
16.	Adjustment Assembly	46.	Meters
17.	Switching Assembly	47.	Transformers
18.	Electronic Switching Assembly	48.	Electronic Software
19.	Bracket Assembly	49.	Planning Bills
20.	Switch Back Assembly	50.	Tools
21.	Differential Rod Assemblies	51.	Electronic Accessories
22.	Housing & Bracket Assemblies	52.	Probes
23.	Knob Assemblies-Mechanical	53.	Series 1 Software
24.	Knob Assemblies-Electronic	54.	Common Parts Sensors
25.	Relay Assemblies-Solid State	55.	Gear Box Assemblies
26.	Relay Assemblies-Mechanical	56.	Enclosure Assemblies
27.	Lead Wire Subassemblies	57.	Printed Circuit Board Assemblies
28.	Miscellaneous Assemblies	58.	Lead Wires-Electronic
29.	Essential Parts-Head Assemblies	59.	Welding Electrodes
30.	Miscellaneous Purchaser Supplied Parts	60.	Nameplates-Mechanical

CLASS CODE DESCRIPTION	CLASS CODE DESCRIPTION
61. Nameplates-Electronic	83. Contacts
62. Flanges	84. Spiral Pins
63. Commercial Fasteners	85. Washers, All Types
64. Plastic Parts	86. Diodes
65. Springs	87. Potentiometers
66. Plastic Stampings	88. Miscellaneous Resistors
67. Springs	89. Prototypes-Mechanical
68. Rolled Tubing	90. Prototypes-Electronic
69. Mineral Insulated Cable	91. Screws
70. Thermistors	92. Valves & Fittings
71. Transistors	93. Contacts, Switch
72. Wiring Diagrams	94. Plastic Moldings & Knobs
73. Special Probes	95. Plastic Tubing
74. Overtravel Assemblies	96. Fixtures, Miscellaneous
75. Adjusting Screw Assemblies	97. Fixtures-Electronic
76. Miscellaneous Actuating Assemblies	98. Fixtures-Mechanical
77. Cover Assemblies	99. Castings
78. Cam Assemblies	100. Castings, Painted
79. Printed Circuit Boards-Auto Insertion	101. Insulating Parts
80. Armored Cable Assemblies	102. Wire
81. Armored Cable	103. Brackets
82. Common Parts--Final Assembly	104. Fittings

* *The number preceding each code description represents a numerical order, not a code—inserted for your convenience.*

As an example, Manufacturing Engineering Manager John Sedgwick recently asked IT for a list of screws in current use in the company. After two weeks of programming and another week of examining the reports, Sedgwick and Denise Andrews, IT Manager, determined that most screws were in Class Code 91—but that others were found variously in: Class Code 38 (switches, parts for), Class Code 33 (miscellaneous commercial parts), and Class Code 63 (commercial fasteners). Figure 7.2 shows a portion of their final report on all active screws at PUI.

Figure 7.2. **Before VEP: Report of Active Screw Part Numbers (Partial List)**

PART NUMBER	CLASS CODE	SCREW DESCRIPTION
6261-258	91	Brass Screw
6261-299	91	Fastener, Screw
6276-091	38	Screw, 1/4-20 Master Switch
6277-001	91	Set Screw
6277-008	33	S/S Screw, Flat Head
6277-035	33	Stock S/S Screw
6277-109	63	S/S Screw
6277-110	91	1" Hex Screw
6277-125	91	Brass Screw
6277-221	38	Screw, Binding 1/4"
6277-231	91	Screw for 2 Series Bracket
6277-485	91	Screw, Same as 125 But S/S
6277-497	91	Commercial Stock Screw
6277-104	91	S/S Hex Head

Both Sedgwick and Andrews suspected there would be similar problems if they used existing class codes to search for other groups of parts (parts types). They also realized they could not use the four-digit prefix <6277> to identify all screws either, because the prefix rule had too many exceptions. They wondered whether this mess could be sorted out—and, if it could, whether there was a way to keep it from recurring.

Trouble Area 4: Description Fields

Each product or part record in the PUI classification system also carries a 30-character alpha-numeric description field. But it is widely known that information loaded into this field is often inconsistent and seldom maintained once it is added. In addition, the descriptions do not provide sufficient information for PUI to make sound decisions relative to new product design. (See Figure 7.3 for examples of the range of 30-character code descriptions in the system.)

Creating A VEP-Capable Classification System

Figure 7.3. **Before VEP: Examples of PUI's 30-Character Codes**

```
M / S   6 - 3 2   X 1 / 4   B / H   S / S

O V E R H A U L   H S G

P R E S S U R E , N O N - I N D I C A T I N G

C O N T R O L   C O M P O N E N T

S C R E W , 1 / 4 - 2 0   M A S T E R   S W I T C H

C O M M E R C I A L   S T O C K   S C R E W

S / S   H E X   H E A D
```

Trouble Area 5: Other Codes

To make matters more complex, the system also carries other codes for parts and products that help determine how they will be planned, manufactured or bought. While these codes are typical of a traditional ERP-based system, they too are not always consistently applied. Therefore, they cannot be considered an effective means for assessing variety or supporting a parts effectiveness process.

IT Manager Andrews is a great supporter of PUI's plans to reduce the number of active parts and processes through VEP. For her group, every new part or process swells an already bulging database. Perhaps more than anywhere else, the IT group recognizes how difficult it is to sort out PUI's voluminous data in any meaningful way. Yet department employees are constantly asked for reports on parts, parts characteristics, parts usage across the product population and so on. Although IT will eventually produce a report, they are quite sure they are not doing it effectively or efficiently.

For the most part, Andrews feels the classification system is adequate for many of her group's purposes—but not adequate for supporting decisions about new parts and products. She also feels her IT Department should and *must* take an active role in making revisions to support variety effectiveness.

Where Group Technology Fits In

Group Technology (GT) is a management system that makes parts variety visible by treating parts-related data as groups rather than as individual, isolated elements. Developed in 1948 by E. G. Brisch & Partners in England, GT was widely used in the decades before JIT, SMED, and lean to reorganize and streamline a company's manufacturing systems.

GT's technology-based process begins with classifying parts with similar attributes and features into families or groups. This is followed by weeding out duplicates, near-duplicates, and other extraneous parts. With that as a base, subsequent tasks are to:

- Identify and analyze the current manufacturing processes (routing sequences) for each parts group.
- Improve routing sequences for all part families.
- Standardize and adopt the most effective routings.
- Utilize these standard routings in all process planning.

VEP and GT are closely allied in several areas. Like VEP, GT recognizes the pivotal role of classification systems in the analysis needed to achieve true and substantial cost savings. Both approaches also acknowledge the challenging, unglamorous work of making these systems capable. Like VEP, GT advocates standardization and prevention. Like VEP, GT benefits from software support.

The difference between VEP and GT lies in the scope of the endeavor. GT focuses on data systems in order to locate and minimize variation so that products can be designed and produced more efficiently and at less cost. While VEP also champions those goals, it widens the inquiry to include the role of every organizational system in triggering and/or preventing negative variety—in all its forms. VEP seeks to build variety effectiveness into every stratum of the enterprise.

As W. F. Hyde, subject matter expert at Brisch, Birn & Partners, put it: "Group Technology is VEP's best friend." If you are currently deploying GT, you are ahead of the variety effectiveness game plan because a lot of the groundwork is bound to be in place. If GT is new to you, then treat it as an option to investigate—but not as a requirement for moving forward. GT is not required for an effective VEP implementation. A sound and capable classification system is!

Making The Data System VEP-Capable

A VEP-capable classification system must contain the data needed to allow engineers to make sound decisions relative to the design of parts and products—and that system must ensure that data can be sorted meaningfully. If not, your system is a barrier to effective variety. Not only is it incapable of solving the problem—but the system itself is actually one of the prime causes of the problem. An overhaul is needed. With so many possible deficiencies and problems in the parts classification system where does one begin the clean-up? What to do—and what to do first?

Begin with Nomenclature

Begin by standardizing the parts terminology. That means: define a single set of terms or names for part types and stick to it. Over time, this nomenclature becomes a common language people use on a regular basis.

A screw, for example, is no longer also referred to as a fastener. Lids are not also covers. Brackets are not also braces and/or angle stays. The point is clear: Without standard names for parts, a company cannot shore up its class codes. And without reliable coding, you cannot move to the next steps.

The benefits of establishing a common nomenclature are immediate and powerful. Figure 7.4 shows the revised set of codes after the Parts Type Team at PUI made terminology improvements. Note that the number of terms was reduced from 104 to 74—a nearly 29% decrease.

Figure 7.4. **After VEP: Active Class Codes at PUI (Reduced from 104 to 74 Class Codes)**

CLASS CODE	CODE DESCRIPTION	CLASS CODE	CODE DESCRIPTION
301	*Over travel Assembly	352	Relay Assembly (Fabricated)
302	Bellows & Press Assembly (Fab)	353	Lead Wire Sub-Assembly
303	Remote Temp Assembly	361	Lead Wire S/A Electronic
304	Local Temperature Assembly	391	Miscellaneous Assembly
307	Bulb Assembly (Fab)	487	Diaphragm Assembly (Fabricated)
308	Adjustable Screw Assembly	651	Welding Electrodes
309	Adjustment Assembly	658	Bellows/Bellows S/A (Purchased)
310	Bellow & Housing Assembly	661	Castings, Enclosures Covers

Smart Simple Design/Reloaded

CLASS CODE	CODE DESCRIPTION	CLASS CODE	CODE DESCRIPTION
311	Actuating Assembly	665	Chart Drive-Electric & SP Wound
312	Switch Bracket Assembly	639	Chart Recorder
314	Cover Assembly	642	Dial plates
315	Switch Bank Assembly	679	Nameplates Decals
316	Enclosure & Base Assembly	692	Valves & Fittings (Mechanical)
318	Chart Plate Assembly	699	Gaskets, Seals (Purchased)
322	Bellows Diff'l Rod Assembly	706	Metal, Rods, Bars, Shapes, Wire
323	Compensator Assembly	713	Misc., Purchased Parts
324	Cam Adjustment Assembly	720	Commercial Fasteners
325	Bellows HS'G & Bracket Assembly	724	Plastic Tubing
328	Pneumatic Valve Assembly	725	Plastic Moldings & Knobs
329	Probe Assembly (Fabricated)	727	Plastic Rods, Bars, Shapes, Sheets
336	Knob Assembly	728	Plastic Clips
337	Printed Circuit Board Assembly	748	Screw Machine Parts
338	Auto Insertion PC Sub-Assembly	755	Screw Machine Parts, Plastic
776	Contacts, Switch	871	Shipping Labels, Boxes, Supplies
779	Separate Wells & Connectors	872	Info/M, Spec, Instruction Sheets.
786	Springs, All Types	910	Capacitors
793	Stampings, Metal	917	Diodes
800	Stampings, Plastic	924	Electronic Components, Misc.
807	Switches, Mesh	927	Potentiometers
814	Switches, Mechanical	931	Printed Circuit Board
821	Commercial Electrical Connectors	935	Probe Housings
828	Thermocouple Components	944	Relays
835	Misc. Metal Tubings	959	Resistors, Point of Use Program
842	Tubing, Plastic	966	Meters
849	Tub, Cap, for Therm. System	973	Thermistors
856	Commercial Insulated Wire	980	Transformers
870	Product Packaging, Individually	987	Transistors

* *The numbers preceding the Code Descriptions represent a numerical order, not a code—inserted for your convenience.*

Second: Define Attribute Templates

Next, decide what kind of information your classification system needs to contain. For VEP purposes, you need to develop a set of parameters that define which part attributes are meaningful. Once defined, these parameters are converted into an attribute template that is used each time a new part enters the database. Termed a *standard attribute template*, we recommend that you adopt this as your only standard and use it to defend against a part being described on an ad hoc basis. Your standard attribute template names the key attributes—for design purposes and for the goals of variety effectiveness.

As we just saw, previous to VEP, PUI described screws in a variety of non-standard ways (see Figure 7.2 earlier). When PUI developed a standardized VEP attribute template on screws, screw specifications were presented in an organized, consistent manner. For example, PUI determined that the telling attributes for screw design were: material, length, threading, and head. Parts numbers are assigned accordingly (Figure 7.5). This template took the confusion out. Product engineers could access the vital information they needed to answer the core VEP question: "Can I use this screw in this new product, instead of introducing yet another new one?"

Figure 7.5. **After VEP: PUI's Attribute Template for Active Screws**

SCREW PART NUMBER	MATERIAL	LENGTH	THREADING	HEAD
6277-001	Brass	1/8	6-32	Flat
6277-002	Stainless	1/4	6-40	Flat
6277-003	Stainless	2	6-40	Hex
6277-004	Brass	2 1/4	8-40	Socket
6277-005	Brass	1	6-32	Binding
6277-006	Stainless	1	6-32	Binding

VEP teaches us that all costs adheres to the part. You see this core principle in action when you build your first VEP Parts Index because it shows you the complexity a single part can trigger in an existing product line. VEP's attribute template is organized around the same principle, unmasking the complexity imprisoned in our incapable classification system.

Steps for Creating Attribute Templates. The often arduous task of creating attribute templates is assigned to the Parts Type Analysis Team; it is their first objective. The team decides on parameters and then designs a template for each parts type, with computer fields designated for each crucial bit of attribute data. In some companies, this means taking values directly off the original drawing and inputting data manually.

If you are thinking that this task can be easily delegated to support staff, think again. Because of the huge likelihood of error when done by non-technical personnel, engineers and technicians are best suited to this job. Given that engineers and technicians do the work themselves, here is the step-by-step procedure they follow:

- The team's starting point is the list of parts types in the targeted series.
- The parts type population in that series is divided among the team's members.
- Each team member defines an attribute template for each of his or her assigned parts types.
- Templates are submitted to the team for review and feedback.
- According to the feedback, each template is improved and, when ready, adopted by consensus.
- Using the template, the team member begins to input the data for the designated parts type, checking drawings directly for accuracy.
- Each team member dedicates a set amount of time to enter data on a regular basis—say, two hours a week or twenty minutes a day.
- A check sheet or scoreboard is maintained to keep team members on track.

Over time, this process is completed for the entire targeted series. Iterations continue through the subsequent items on the VEP product priority list until attribute templates for *all* part types are entered.

The previous illustration on screws is a simple one, and yet impressive economies were achieved. Imagine the multiple impacts on parts inventories and system complexity when applied to all commodity parts— as just one example. When the time comes, this clean-up process enables the Parts Type Analysis Team to ask and answer its central questions:

- Do we really need this part?
- Do we really need this particular variation of this part?
- Can we eliminate, combine or substitute this part in some way?

Third: Tackle the Class Codes

In parallel with developing standard attribute templates, the company sets and maintains clear, reliable data input procedures. Assigning class codes, for example, is no longer a matter of expedience or personal preference.

The PUI Parts Type Team was made responsible for upgrading all procedures related to the classification system. Team members worked hard to develop good guidelines for determining class codes for parts. Part of that process—and of ensuring widespread buy-in—was getting ideas and feedback from all concerned parties. You can bet that the Drafting Department, as one of the main recipients of uncontrolled parts variations, had plenty of suggestions. As shown in Figure 7.6, PUI came up with a set of simple guidelines that serve as the base for further systems upgrades.

Figure 7.6. **Class Code Guidelines for Parts**

	CLASS CODE GUIDELINES FOR PARTS
1.	All parts within a Class Code will carry the same prefix numbers in the part number designation. For example, all screws begin with 6277, followed by a dash (-).
2.	Numbers following the dash (-) in a prefix designator are assigned sequentially when new part numbers are required.
3.	Any new part is assigned to only one approved, existing Class Code, based on the most current Class Code list.
4.	Each new part number is assigned to the existing Class Code category that carries all similar parts. If such a class code does not exist (or if there is any question whether a part fits within a certain Class Code), the VEP Parts Type Team Leader (together with the Drafting Manager) will determine how to classify the part.
5.	Pre-defined attributes for each Class Code shall be the deciding factor in determining if a particular part fits within a particular Class Code.
6.	The Parts Type Team Leader (or the Drafting Manager) must review and approve the coding of all new parts before they are entered into either the computer system or the attribute database.
7.	The Parts Type Team Leader (or Drafting Manager) must approve any requests for Class Code changes in an existing part.
8.	A Class Code shall not have overlapping or redundant parts unless special agreement is reached with the Parts Type Team Leader or the Drafting Manager.

This completes our treatment of the three-part VEP procedure for improving your parts classification system so that it is capable of supporting VEP decision making. Doing this harnesses the power of the data in your IT systems so that information can join in the battle against high parts inventories and negative variety.

With a VEP-capable classification system in hand, we now turn to the set of sturdy analytical tools for the next step in the VEP Methodology: The Six VATs.

CHAPTER 8

The Six VAT Tools

> If a hammer is your only tool,
> all problems begin to look like nails.
>
> — ANONYMOUS

With your classification system now in an improved state and your team structure in place, you are ready for Stage 2—and the action in Stage 2 is reduction analysis.

Begin your search for reduction opportunities by identifying negative variety in three key areas: market offerings, the structure of products, and the scope and composition of the company's parts universe. Shorthand for this in VEP is *3-View Analysis*, with three main teams: Market Analysis, Product Structure Analysis, and Parts Type Analysis; these are discussed in detail in Chapter Nine. Two other teams—Process Analysis and Control Point Analysis—are optional and discussed in Chapter Ten.

The core set of tools used by all these teams is called: *The Six VATs*. Explaining what they are is the focus of this chapter (with thanks to Toshio Suzue and Akira Kohdate, *Variety Reduction Program*, and Geoffrey Boothroyd and Peter Dewhurst, *Product Design for Assembly*).

The Six VATS: Tools Of Inquiry

The Six VEP Analysis Tools (VATs) are:

VAT-1	Unique vs. Shared	VAT-4	Ease of Assembly
VAT-2	Modularity	VAT-5	Range
VAT-3	Multi-functionality & Synthesis	VAT-6	Trend

All of the 3-View Teams use the VAT tools to varying extents and in varying combination, depending on their point of inquiry.

The Product Structure Analysis Team, for example, makes extensive use of VATs 2 through 4, while the Parts Type Analysis Team relies heavily on VATs 1, 5, and 6. The Market Analysis Team, on the other hand, applies the *principles* behind all the VATs (rather than the engineering tools themselves) to diagnose and then rationalize customer and market decisions related to the company's product and service offerings.

VAT-1: Unique vs. Shared

VAT-1, the first of the six tools, begins with the question: Is this part/product/offering *shared* (used across several or many products lines)—or is it *unique* (dedicated to one specific product, exclusively)? The principle of unique vs. shared is applicable to all analysis targets: marketing, product structure, parts—as well as processes and control points. Do not be deceived into limiting the use of VAT-1 to parts, simply because we use the following parts-based example. Ask yourself:

- Is this part *unique* to this model—or shared with another model or several models?
- If the part is shared, can it be shared more widely? That is, can it be further commonized or standardized to serve additional product structures?
- If the part is unique (also known as dedicated or non-standard), the question becomes: Why is it unique? And, can we standardize the specifications sufficiently so that it can be shared with at least one other model?
- If the part is already shared, we apply the other five VATs in order to determine if it can be further commonized.

On the other hand, if the part is dedicated or unique, we need to determine if the reason for this "specialness" is valid. Is it a dedicated part because of a specific customer demand? Or is it the result of an inefficiency or anomaly in the company's internal practices? In other words, we need to determine if the variation is customer-driven or internally-triggered.

If the variation is verifiably customer-driven, the difference constitutes *positive* variety and represents an important profit advantage. If, on the other hand, the variation is internally-triggered—caused (however unintentionally) by a practice within the company—it represents *negative* variety and is a suspect cost. This line of logic remains true even for product design requirements related to form, fit, or function.

Yes, we mean it! In VEP, *even* variations caused by the requirements of form, fit, and function are classified as *negative variety*—if these variations are not specifically requested by the customer. There may be no known way around these requirements. Your product engineers may be forced to spec-out the part in a way that renders it unique in order to fulfill the demands of the product architecture or because of production constraints. In VEP, however, we want to be fully aware of that. We want to understand *that* we are creating negative variety—even if there is no way around it, at least not at the moment. It still counts as negative because it is internally-triggered.

The work of VAT-1 is to locate negative variation in all things: the structure of products, the way products are segmented for market, the specifications of the parts themselves—no exceptions, no exemptions. Every VEP Team seeks to shift as many unique or non-standard elements to a shared status as possible—or to understand why they cannot.

But there are limits. For example, there may be limits to the amount of parts sharing across certain product lines. Some products are specifically designed to contain one or more variable parts, expressly to make them more responsive to the marketplace. This is the part of the product that changes when, for example, mauve replaces purple as the color of the year—or ships replace alligators as insignia of distinction.

Some industries have special constraints. A supplier that globally sources truck interiors to the Big Three automakers is obliged to design products—a distinctive visor and innovative cup-holder—that make each of its customers stand out in the marketplace. Because these customers are tooth-and-nail competitors, unique distinctions between product lines must be preserved; they are not candidates for commonization.

But what about the infrastructure, the internal workings of these "unique" products? Returning to the discussion on Nissan and the variety explosion of the 1980s (Chapter Five), we ask if a product's fasteners, hinges or O-rings need to be as distinctive and exciting as the surface of the product? On the flip side, a customer's request for a special element may not necessarily preclude its being used in another product, as a crossover. Teams with VEP training get good at finding such opportunities. When the first crossover opportunity is found, it is a breakthrough. Later, it becomes routine and expected.

The Benefits of Sharing

Unique parts in a product can constitute a significant competitive advantage when they result in product features that make the company's products distinctive in the marketplace. But variation at the parts and assembly level is rarely required or specifically desired by the customer. When that is the case, the variety is wasted. It adds complexity and costs, with no payoff.

There are other costly ramifications. Production reference points on non-standard parts often vary and can cause wide fluctuations in production processes. When these reference points change from model to model, the need for special fixtures and processing escalates—as does the need for differing assembly techniques. If, on the other hand, parts are sufficiently standardized, a single standard production process can be used for multiple parts, with no added fixtures, changeovers or adjustments.

Clearly, converting dedicated parts into shared parts can result in considerable cost savings related to control points as well. Since shared parts mean fewer parts, layers of transactions—such as parts reordering, material handling, incoming inspection, and counting—are automatically reduced every time the ratio of shared over unique parts increases.

Figure 8.1 shows the results of one company's assessment of parts across all of its product lines (series), using the unique vs. shared yardstick. Those results spoke volumes about the levels of costly complexity caused by unplanned proliferation. Most importantly, the VEP Parts Type Team—and subsequently the Steering Team—recognized the urgent need to address the costly proportion of unique parts, as columns D and G so amply demonstrate. Notice as well that the ranking in this chart's first column is entirely focused on the number of unique/dedicated parts in each series and across all series.

For VAT-1 to really work its magic, all parts must be examined—on the model and product series levels. No matter how low its intrinsic value, let no individual part escape scrutiny. When a so-called "unique" element is found, question its status: Must it remain unique? Why?

Taken as a whole, this first VAT tool (Unique vs. Shared) forces the company to assess its current level of standardization across all variety-triggered functions—marketing, product structure, parts, control points, and production processes—and ask: Can the level of standardizatio n be expanded? In this way, VAT-1 is the first key for maximizing resources and

Figure 8.1. **Unique vs. Shared Parts Index: Series Level**

Rank	Series	A Unique Parts	B Shared Parts	C Total Parts	D Number of Shipped Units	E (A x D) Parts Index of Unique Parts	F (B x D) Parts Index of Shared Parts	G (E + F) Total Parts Index
1	54	273	304	577	65,343	17,838,639	19,864,272	37,702,911
2	120	282	694	976	32,625	9,200,250	22,641,750	31,842,000
3	400	236	521	757	16,854	3,776,000	8,780,934	12,556,934
4	55	205	206	411	15,729	3,224,445	3,240,174	6,464,619
5	40	79	126	205	39,841	3,147,439	5,019,966	8,167,405
6	800	402	353	755	6,903	2,775,006	2,436,759	5,211,765
7	650	431	303	734	4,752	2,048,112	1,439,856	3,487,968
8	930	233	268	501	6,674	1,555,042	1,788,632	3,343,674
9	6	81	282	363	15,047	1,218,807	4,243,254	5,462,061
10	100	62	440	502	20,089	1,245,518	8,839,160	10,084,678
11	59	48	102	150	12,725	610,800	1,297,950	1,908,750
12	21	54	110	164	9,521	514,134	1,047,310	1,561,444
13	720	172	36	208	2,822	485,384	101,592	586,976
14	920	229	302	531	2,476	567,004	747,752	1,314,756
15	105	71	322	393	6,094	432,674	1,962,268	2,394,942
16	950	115	198	313	3,570	410,550	706,860	1,117,410
17	41	70	102	172	4,187	293,090	427,074	720,164
18	4	54	168	222	4,792	258,768	805,056	1,063,824
19	970	168	180	348	1,351	226,968	243,180	470,148
20	35	104	90	194	1,680	174,720	151,200	325,920
21	119	87	163	250	1,885	163,995	307,255	471,250
22	8	13	1	14	6,119	79,547	6,119	85,666
23	117	24	354	378	3,300	79,200	1,168,200	1,247,400
24	680	172	185	357	393	67,596	72,705	140,301
25	24	30	14	44	1,616	48,480	22,624	71,104
26	10	31	14	45	1,113	34,503	15,582	50,085
27	820	33	299	332	803	26,499	240,097	266,596
28	459	20	144	164	381	7,620	54,864	62,484
		3,779	6,281	10,060	288,685	50,510,790	87,672,445	138,183,235

minimizing costs throughout the enterprise. Plus it paves the way for applying the other five techniques in the VAT tool kit. For a summary of VAT-1, plus a simple procedure to guide the inquiry, see Figure 8.2.

Figure 8.2. **Summary of VAT-1: Unique vs. Shared**

DEFINITIONS	**Shared Parts** are parts that have been sufficiently standardized so they can be used—shared—across several or many models. Also known as commonized parts. **Unique Parts** have not been standardized or shared either because they are linked to a special customer demand (positive) or are constrained by technological/organizational limitations (negative)—or both. Certain previously unique parts, on closer look, may lend themselves to commonization.
OBJECTIVES	• To separate the constituent components of a product (or product offering) into those that have been commonized or standardized—from those that are unique or non-standard. • To validate or refute a component's unique status. • To widen the standardization range on existing shared components—and convert as many dedicated (unique) ones as possible to shared status.
KEY QUESTIONS	• What is the current level of standardization? • Can it be increased? • Part-by-part, how valid is the premise that customer requirements preclude commonization? • Which parts should not change from model to model and why?
APPLICATION PROCEDURE	**Step 1.** Separate the constituent parts of a product series or family into elements that are unique (dedicated) and those that are shared (standard). **Step 2.** Widen the scope of each shared part so that more models are covered—then commonize as many currently dedicated elements as possible. **Step 3.** Develop and adhere to guidelines that promote the commonization of parts—and stipulate clearly when parts must remain unique.

VAT-2: Modularity

VAT-2, the second of VEP's six analytic tools, is *Modularity*. As with VAT-1, the principle of modularity is applicable to all VEP analytics, whether marketing, product structure, parts, production processes or control points.

Let's return to our product-focused example. Modularity for product development shifts the perspective from individual and isolated parts to parts in groups or subassemblies. The part is now connected with its place in the product environment or architecture.

In this second VEP tool, the relationship parts have to each other within a product takes over: how parts mate. (Later the focus moves across products.) In Figure 8.3, the four parts to the left give us the VAT-1 perspective: distinct, separate, and isolated from each other. On the right are the same four parts from the perspective of VAT-2: as a unit or subassembly in relationship to each other.

VAT-2 forges the link between standardization and interchangeability by challenging us in three ways:

- To what extent can subassemblies (or given sets of parts) be standardized into units or modules that can be used in other products?
- To what extent can assorted modules be further standardized so they are interchangeable across an expanded number of products?
- To what extent can product structures be developed so they can accommodate standard modules—or a wide assortment of such modules?

Capitalizing on the increased levels of commonized parts resulting from VAT-1: Unique vs. Shared, modularity seeks to extend the concept of standardization in terms of both functions and dimensions. In this way, parts-mating and module-mating across products are increased. A modular approach also encourages changes in design specifications that do not necessarily require associated changes in existing product structures. New designs are achieved simply by introducing new combinations of modules. Because units are interchangeable, they result in a wider variety of finished products merely by switching them.

Combining modules in this way is a simpler and more economical means for meeting the market's demand for "new" products than creating a product from scratch. An added benefit is that modularized units are easier to upgrade—or downgrade—in response to shifts in market forces. In one company, application of VAT-2 alone resulted in a 30% reduction in parts. When implemented broadly, modularity can trigger powerful simplifications in a company's production system, making this tool even more attractive as a cost-savings technique.

Figure 8.3. **VAT-1 and VAT-2: Two Perspectives on the Same Parts**

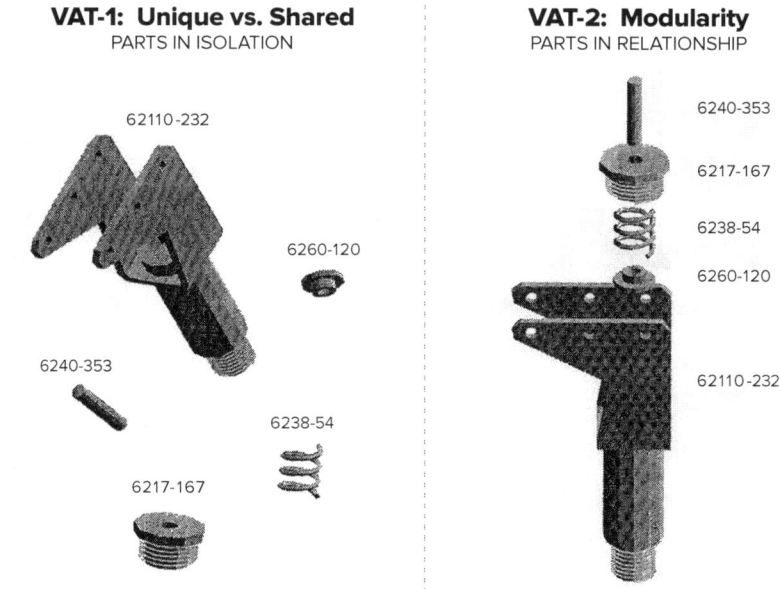

Three Modular Styles

There are three primary ways to achieve modularity in products: amplification, replication, and combination.

Amplification. Amplification is an approach that requires a stable product base or core. As parts are added to the base, the variety of new products expands.

The children's toy Mr. Potato Head works on this basis. An array of plastic parts (noses, eyes, ears, hats) is added to the product core—the potato itself. The result is a host of weird-looking, impressively distinctive creatures. The same concept is used in making pizza pies. The base is constant: the crust. Then all kinds of ingredients (parts) can be added, making potential combinations and the possibility of satisfying customers practically unlimited.

In another universe, the Prizm and Metro models of GM's now-defunct Geo line shared an identical base—the same chassis, engine, and power train. Models diverged in terms of seats, interiors, trunk styles, dashboards, and so forth to expand the range of appeal. The same approach is popular in footwear, especially in sneakers.

Replication. The replication approach uses a series of standardized, identical modules that result in different products depending on the number of modules used.

This is exactly the way children get so many fantastic creations out of a set of plain wooden blocks or plastic Legos®. The components are the same, but because they share common reference or mating points, they can be put together so that different end-items result. Stackable office or kitchen trays and CD racks are two other examples of replication.

Combination. The combination approach seeks to create diverse products by linking up standardized modules of varying functions. This is similar to the way children can make a range of "things" by combining different elements in their Erector Sets—cross beams, motors, winches, buckets, corners, roofs, etc. The result is a seemingly endless stream of highly differentiated constructions, limited only by the imagination. The key again lies in the high level of standardization in mating dimensions and reference points. Modularized shelving units are a good example of the combination approach.

VAT-2: Modularity is summarized in Figure 8.4.

Figure 8.4. **Summary of VAT-2: Modularity**

DEFINITIONS	**Modular Design.** A unit or group of standardized elements or parts that may be used within a number of different products because the unit itself is sufficiently commonized to be interchangeable across a number of products.
OBJECTIVES	• To minimize the number of required elements by combining as many as possible into standardized modules that are interchangeable. • To create units of exchangeable elements (modules) in order to create wider customer choice while minimizing cross-product complexity and costs.
KEY QUESTIONS	• What is the current level of modularity in products? • How can it be increased? • Can a given standard module be combined with another to create further augmented, replicated or combined models? • Can other modules be combined or re-combined to serve a wider range of function?
APPLICATION PROCEDURE	**Step 1.** Examine a model's assorted shared elements, looking for opportunities to combine these further into standardized modules. **Step 2.** Look across products for opportunities to replace highly variant sub-assemblies with these standard modules. **Step 3.** Develop and adhere to guidelines that promote modularity in new product design.

V-Costs and the First Two VATs

These first two VATs are aimed primarily at minimizing the scope of variation, whether across the marketing, product structure or parts analysis view. When applied by the Parts Type Team, for example, successful results can dramatically lower V-Costs (Variety Costs/Chapter Three), including secondary variations in production processes, process paths, equipment, fixtures, dies, labor hours, and control points.

VAT-3: Multi-Functionality & Synthesis

VAT-3: Multi-functionality & Synthesis concentrates on reducing F-Costs (Function Costs) by creating parts and units that serve multiple uses and perform multiple functions, thereby reducing the number of parts required to fulfill those functions.

Closely linked to the principles of Design for Manufacturability (DFM), the *multi-functionality* facet of VAT-3 directs engineers to design for robustness so that a greater range of function is served through the same or fewer parts. The second element of this VAT—*synthesis*—focuses on finding new materials and new engineering technologies that allow previously separated parts to be merged or collapsed. As a result, the same or additional functions are met through fortified but less variant specifications. The goal is product designs that keep to an absolute minimum the number of constituent parts needed to fulfill each product function.

The analysis starts, as always, with the part. Here we determine if a given part is suitable for a VAT-3 application. To do this, think of the part in its context—as an element in the structure of the product—and answer three questions:

- Does the part have to move separately from the other parts during the operation of the product?
- Does the part have to be of a material different from that of other product parts?
- Does the part have to be removed for servicing or re-assembly, separately from the other product parts?

If the answer to *any* of these questions is "yes," the function of this part cannot be merged with that of another. If the answer to *all* three questions is "no," the part is a possible candidate for multi-functionality and synthesis and you are ready for the next steps.

- Identify the product's required functions.
- Understand how these functions are fulfilled through the constituent parts.
- Search for ways to simplify, integrate, substitute or optimize parts and functions.

Your solutions may take several directions. First, you may eliminate or remove extraneous specifications or surplus functions. In the case of the bracket function shown in Figure 8.5, for example, four parts and at least two kinds of material were required—a low-carbon steel bracket, stainless-steel fingers, and two fasteners. Later, the two components are combined and both fasteners eliminated.

Please Note: The examples in the remainder of this chapter are very simple for a reason: so that any reader unfamiliar with the techniques under discussion can readily understand the underlying principles.

Figure 8.5. VAT-3: Multi-Functionality & Synthesis: Bracket (Before/After)

Alternatively, you can combine or incorporate several functions into one. The design of a pressure control, for example, called for two separate parts. One part had the function of guiding the load spring that changes the range of control. The second part was the plunger, a rod that transfers the movement from the sensor to the electrical switch. The two components did not move with respect to each other; they rode on top of each other. When VAT-3 was applied, it was clear that the two functions could be merged. The result was the single, multi-functional part shown in Figure 8.6.

Applications of VAT-3 have an exceptional potential for reducing parts and processes. Regardless of its low intrinsic value, every single constituent part must be queried. One company estimated that the elimination of a single fastener would save $15,000 over the life of a redesigned point-of-sale terminal.

Figure 8.6. **VAT-3: Multi-Functionality & Synthesis: Plunger (Before/After)**

While a VEP team may eventually determine that a part cannot be eliminated because, for example, the required specialized equipment to manufacture a multi-functional version of the part is not available—or due to other diseconomies of manufacture. But such constraints are discovered only *after* VAT-3 is applied. See Figure 8.7 for a summary of VAT-3: Multi-functionality & Synthesis.

Figure 8.7. **Summary of VAT-3: Multi-Functionality & Synthesis**

DEFINITIONS	**Multi-functionality** seeks to design for robustness, formulating product structures that include only the minimum number of elements to fulfill required functions, with each part serving a greater range of specification. **Synthesis** looks for ways to combine, integrate or minimize such parts further by using new materials, production technologies or structural concepts that use fewer parts to meet required functions.
OBJECTIVES	• To minimize the number of elements required to fulfill specific functions by making each element serve as many different functions as possible. • To create a product that provides the desired functions through a simplified structure by eliminating differing materials and making the remaining materials more robust.
KEY QUESTIONS	• Can a function of this product or part be merged with that of another? • If so, can the same level of capability and customer selection be maintained or expanded?
PROCEDURE	**Step 1.** Identify specific functions of the part, components or product. **Step 2.** Look for ways to meet these functions through a simplified product structure, requiring fewer parts, components or sub-assemblies. **Step 3.** Look for ways to minimize the number of parts further by using different materials, production technologies or structural concepts. **Step 4.** Develop and adhere to guidelines that promote multi-functionality and synthesis in new product design.

VAT-4: Ease Of Assembly

The first three VATs are aimed at reducing complexity and its attendant costs through variety reduction and design simplification. In VAT-1, we standardized as many elements as possible. In VAT-2, we combined standardized elements into modular configurations. In VAT-3, we integrated functions and material requirements into a multi-purpose result. Only essential elements remain. Now it is time to ensure that these elements are easy to configure or assemble.

This is the function of VAT-4: Ease of Assembly. Some of the important guidelines for accomplishing this include:

- Eliminate or simplify adjustments.
- Minimize the use of separate connectors and fasteners (see Figure 8.5 above).
- Design parts that are self-aligning and self-locating (Figure 8.8).
- Design parts that cannot be installed incorrectly (Figure 8.9).
- Ensure adequate access and unrestricted vision during assembly (Figure 8.10).
- Ensure ease of handling of parts from bulk (avoiding parts that nest or tangle, are flexible, fragile, sticky, magnetic, abrasive, too small, too light, or too big).
- Minimize the need for reorientations during assembly; design for top-down assembly.
- Maximize part symmetry or make parts obviously asymmetrical.

We are grateful to Geoffrey Boothroyd and Peter Dewhurst for their permission to use examples 8.5 through 8.11 from their book, *Product Design for Manufacture and Assembly*.

Figure 8.8. **VAT-4: Ease of Assembly: Spring (Design Self-Aligning/Self-Locating Parts)**

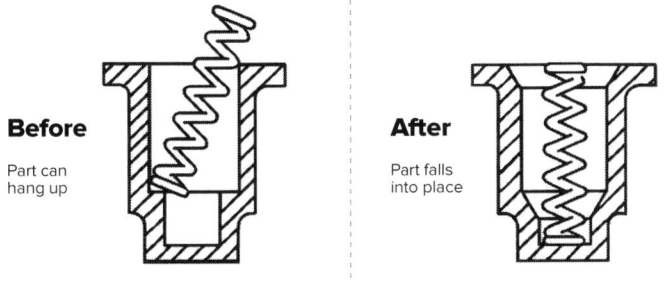

Before
Part can hang up

After
Part falls into place

Figure 8.9. **VAT-4: Ease of Assembly-Pin (Parts Cannot Be Installed Incorrectly)**

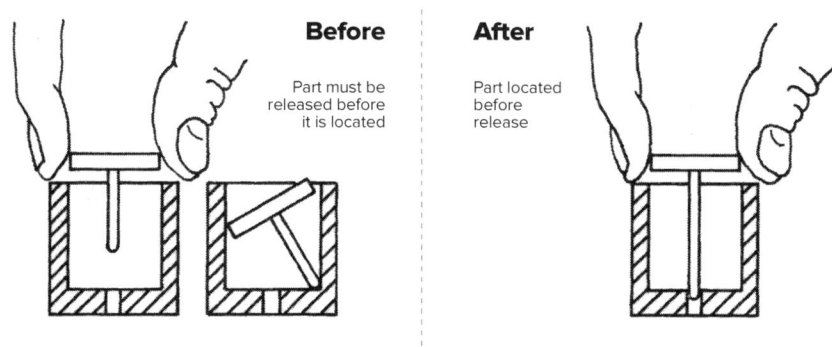

Figure 8.10. **VAT-4: Ease of Assembly: Housing (Adequate Access/Unrestricted Vision)**

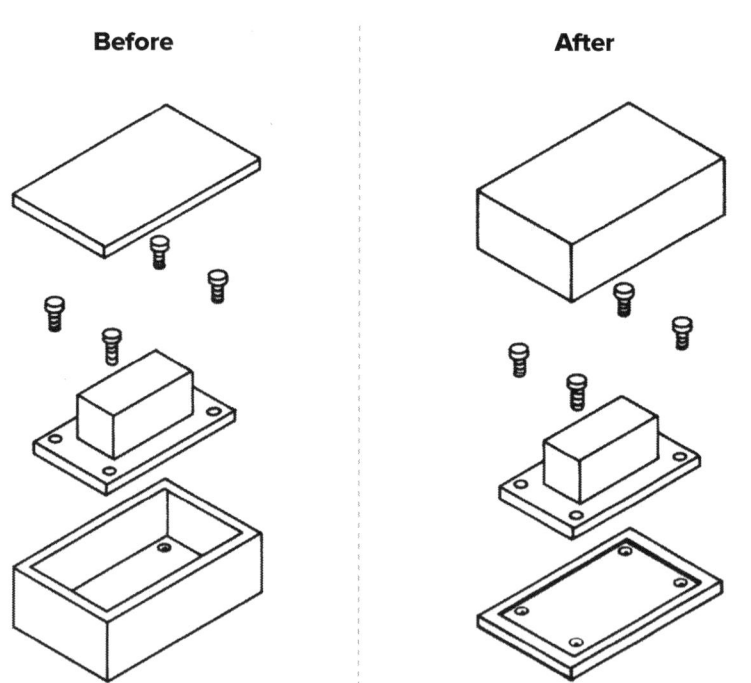

Because assembly costs are largely set during the product engineering stage, it is paramount that engineers work with ease-of-assembly considerations from the outset of the product's design (Figure 8.11).

Figure 8.11. **Summary of VAT-4: Ease of Assembly**

DEFINITIONS	**Ease of Assembly** means that, after parts have been upgraded through applying VATs 1 through 3, we analyze all constituent parts to ensure that they are easy to assemble.
OBJECTIVES	• To make all remaining parts as easy as possible to assemble. • To improve costs associated with the assembly process by creating a product structure made of parts and components that are easy to handle and assemble.
KEY QUESTION	• Can part be assembled from above? • Can part be made to require few or no adjustments? • Can part be made requiring no or minimal types of fastening? • Can assembly motions requiring skills be minimized or eliminated? • Can differences between like parts be made obvious?
PROCEDURE	**Step 1.** Dis-assemble the product, part by part. **Step 2.** Re-assemble the product, part by part, using VAT-4 assembly guidelines and generate alternate part concepts. **Step 3.** Develop and adhere to guidelines that promote ease of assembly in new product design.

VAT-5: Range

VAT-5: Range and VAT-6: Trend are the last two techniques in the VAT tool kit. Together, they take the broadest possible perspective on parts specification, product structure, and product line because they target the accumulation and pattern of variation within and across a company's entire product and parts universe. Let's learn how VEP utilizes range.

Widely used in statistics, *range* is the simplest measure of the dispersion of a group of values that share some common observable characteristic. Range is the difference between the largest and smallest of those values, including all the values along the way. In VEP, the range technique is used to identify the reach or boundary of variation relative to parts, products, and product lines. Let's consider parts first.

Companies with no unifying product development approach often experience a mushrooming of parts specifications that parallels expanding product variety. This can be further intensified by the lack of a VEP-capable classification system. Given this, product engineers are hard-pressed to locate existing parts for immediate use—or that could qualify for a simple redesign of the product geometry. As a result, new parts come flying into the system. *No one notices because no one has been designated to notice.* There is nothing in place to prevent it. As a result, the range of parts values (among other things) inevitably widens. Similar but not identical parts continue to accumulate, and redundancies and overlaps become numerous.

We begin VAT-5: Range by selecting a specific attribute variation within a specific parts type. Lead wires, for example, share variant attributes such as gauge, material, and color. Springs, as another case, share such variable attributes as *spring rate* (extent to which the spring length changes when force is applied), *outer diameter* (O.D.), and *inner diameter* (I.D.). The range technique enables us to see the spread or dispersion of alterable values, in this case within a specific part type. In this application of range, we are not identifying variation generated on the production floor—but, instead, simply the variety within an attribute of that part within a parts type. Let's look at springs at PUI.

Figure 8.12 shows the dispersion of O.D. values within PUI's spring parts type—PUI had 80. When the Parts Type Team saw this spread, they immediately noticed a dense central cluster of values with a few scattered along edges. When they realized that the dense center clustered around a few frequent values within a very narrow range, the team instantly saw the possibility both of collapsing those clusters and of extending others to accommodate more variety across fewer O.D. sizes. They also knew that before they could recommend these changes, they would have to determine which, if any, of those current values were required by the customer.

Figure 8.12. **VAT-5: Range: O.D. Values in a Scatter Diagram**

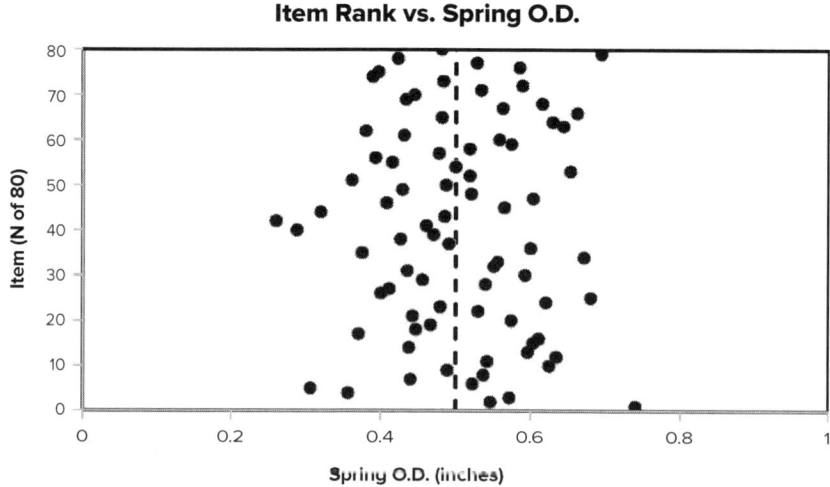

In a different company, the Parts Type Team—also looking at spring O.D.s—derived their insights into range by using a histogram (Figure 8.13). They saw they could standardize diameters within each range and dramatically reduce the variety of spring parts, favoring the less costly standard purchased springs and avoiding costly custom springs. They then identified non-overlapping O.D. ranges for each of their most frequent spring diameters. The bars across the top of each set of values in this histogram separates the spring ranges into viable clusters for possible compression, indicated by the tall, dotted line boxes.

Figure 8.13. **VAT-5: Range: O.D. Values in a Histogram**

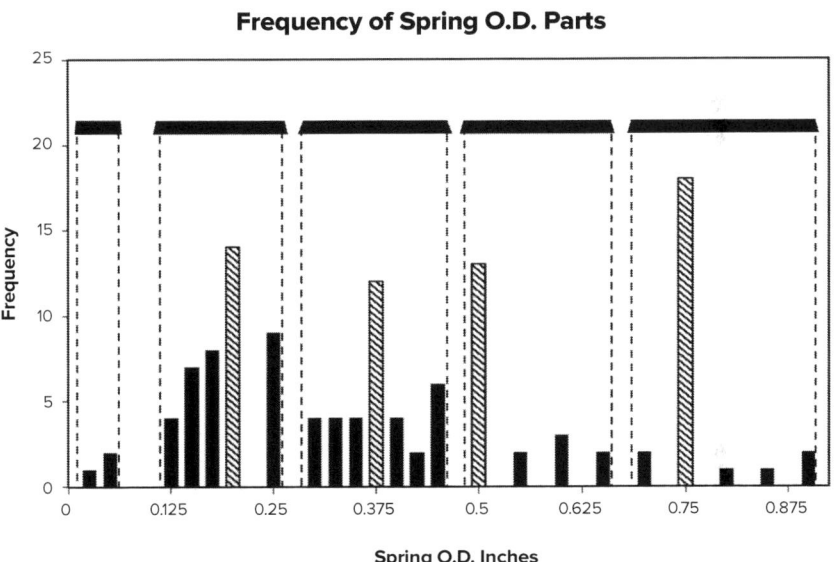

This line of inquiry is exactly the point of VAT-5. Observing the configuration of values in a range dispersion—especially ones with high contrasts (as seen above)—prompts us to investigate the origins of these variations. When we take this on, the other VATs are then brought into play and applied, value by value, and reduction opportunities are identified.

The search for range—and later trend—reduction opportunities requires VEP teams to compile and digest a good deal of numerical information. Both these techniques present data in a useable and meaningful form so an overall picture can be detected. Because range and trend use aggregate data, teams can construct graphs, tables, and figures that visualize results and make relationships between differing values distinct. For example, when product engineers examine parts types in their full value range, they discover overlaps and opportunities to collapse parts, sometimes as many as five into one. A set of convenient standard values emerges that guides the further commonization of parts. For a summary of VAT-5: Range, see Figure 8.14.

Figure 8.14. **Summary of VAT-5: Range**

DEFINITION	**Range** is a mathematical measure of dispersion that shows values that share some common observable characteristic, displaying the difference between the largest and smallest value and the points along the way.
OBJECTIVES	• To organize attributes and other values in order to determine the degree of variation or overlap among those values. • To minimize the number of values needed to cover the range that customers require.
KEY QUESTIONS	• What attribute range width is required for a part or parts across all products? • Do the specifications of certain products or parts overlap or repeat? • Can certain values be merged with others so that the total number of required parts is reduced?
PROCEDURE	**Step 1.** Compute the values range for all specifications within a given parts type or product grouping. **Step 2.** Determine minimum and maximum range widths for each parts type and identify any overlapping or extraneous values. **Step 3.** Eliminate any overlaps or extraneous values by defining a dimensional range, wide enough to encompass the required values but with minimum part numbers. **Step 4.** Develop and adhere to guidelines that support the range technique.

Other Applications of Range

While the range tool has obvious applications in parts type analysis, it is also effective in other analytical aspects of VEP. Market analysis is a good case in point. Marketing departments often trigger requests for new products that are similar but not identical to existing ones. This may be a legitimate tactic, linked to the company's improvement-driven new product approach. But, such requests are often not linked to a unified plan and, as a result, can introduce product variety that is, in fact, negative variety.

As previously discussed, such variations often masquerade as customer-driven requirements; in reality, they are internally-triggered and therefore not valid. The net effect is product overlap and redundancy in market offerings. This issue rests at the core of the 3,200 different snack food products, mentioned earlier, that the now-defunct Borden Foods once offered.

For those motivated to look, VAT-5: Range can help companies identify the tendencies, penchants, pets, anomalies, compensating behaviors, and other biases in their decision-making relative to the products they design, make, and sell. An impressive number of insights and economies are achievable when the range of product characteristics is exposed to the scrutiny of a team that understands the principles of variety effectiveness.

VAT-6: Trend

Just as specifications can span a range of values over time, these same values trace an observable pattern when plotted as a group. The sixth and final analytical technique (VAT-6) is called *Trend*. Its purpose is to help us identify the nature and direction of our business choices, based on the values that dominate in the three spheres: market, product, and parts. This positions us to ask and answer: Is this what we intend? Is this what we need?

In marketing, for example, do most products fall into a high-profit category? Is there an emerging tendency to concentrate on low-end offerings? Is the company as a whole drifting toward commercial business and away from the individual consumer? What is the mix of commercial vs. consumer business over the last five years?

In product design, is the number of parts per product on the rise? At what rate? At what curve? Are there discernible trends supporting more modular designs for miniaturization? In parts, do the data show a growing bias for less or more durable materials? Are outer diameters widening or shrinking? Do choices in switch housing incline more toward explosion-proof or fire-proof?

Trend data can be a powerful tool for revealing issues related to variety effectiveness. In one company, for example, the pattern of trend data on new product offerings revealed a positive drift (the line tracing the center of the trend data). In response, the company developed ways to affirm that direction, consolidate the bias, and strengthen the overall pattern. In another organization, where trend data on purchased parts indicated problems, teams reviewed departmental practices and took steps to arrest the drift in favor of a more advantageous core pattern.

Trend data can also provide important substantiating information. This was the case when the Parts Type Team spotted an anomaly across several component groups: Many new parts entering the system were getting dimensionally smaller and smaller. The team asked the Market Analysis Team for an opinion. The market team generated a trend chart (Figure 8.15) that linked decreasing size to growing requests for miniaturization from OEMs (original equipment manufacturers). They had reduced the size of their products—and so they needed smaller components as well.

Figure 8.15. **VAT-6: Trend in Requests for Miniaturization (Six-Month Intervals)**

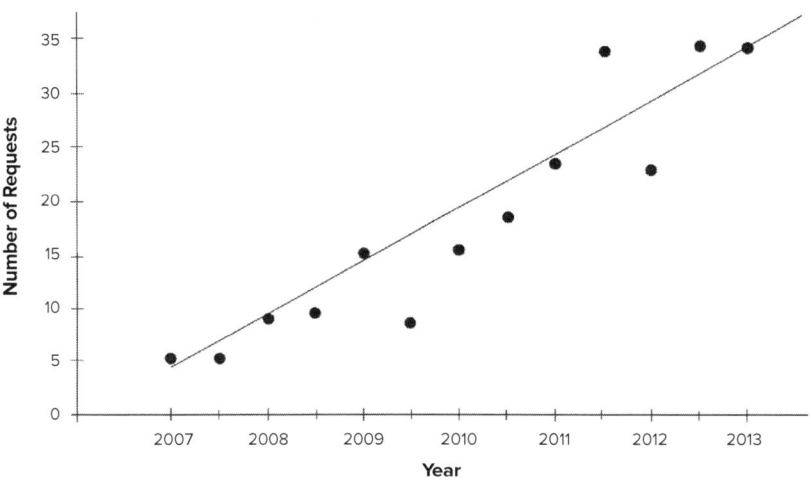

Figure 8.16. **Summary of VAT-6: Trend**

DEFINITION	**Trend** is the discernible pattern and direction of a group of values, sharing some or several characteristics in common when laid out in some pre-set order (e.g., ascending value, descending value, and so on).
OBJECTIVES	• To display and identify the drift in attribute values, specifications, and characteristics as they accumulate in other discrete segments (such as time, market segment, and season) in order to determine a normalized center of such variation so that the number of parts can be minimized. • To determine which drifts support and which drifts obstruct the principles of positive variety.

The Six VAT Tools

KEY QUESTIONS	• What is the pattern of the spread of the company's design, product, and parts choices—in measurable terms? • What is the genesis of that pattern? What did it result from? • Does it support positive variety and the company's strategic direction? • Do we want that pattern to continue?
PROCEDURE	**Step 1.** Study the VEP database for patterns in both the performance-related and dimensional specifications of selected products or parts. **Step 2.** Organize the values of each pattern statistically for the purpose of determining any ordering principles within it. **Step 3.** Plot these values as a numerical trend. **Step 4.** Apply a single consistent principle relative to numerical values and use this to standardize against.

Trend data also apply in the area of corporate and departmental policies where they help us understand that negative patterns form over time and that organizational complexity is not an overnight phenomenon. Figure 8.16 summarizes VAT-6.

The Power Of The Six VATs

While we can become experts at identifying the presence of negative variety, what do we do about it? VEP's six analytical tools (Six VATs) allow us to pinpoint the precise location of negative variety and develop viable proposals for reducing or eliminating it.

The tools do this by helping teams scrutinize the dimensional and functional values of the elements under consideration. Without these tools, the detailed differences between values and the way they are structured can escape detection—whether found in a marketing array, a series of products or the specifications of parts. The Six VATs provide a way to search out variation, based on the questions each tool targets. Here they are within a product/part framework.

Can we share it? If not, how is it different?

VAT-1: Unique vs. Shared

Must this difference remain? If so, can it be standardized enough to share it with at least one other subassembly or model?

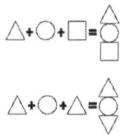
VAT-2: Modularity

Can standard portions of this product be further standardized so they become building blocks for many other products?

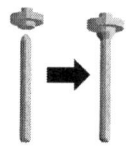
VAT-3: Multi-functionality & Synthesis

How many functions can a single part serve? Can this be extended even further by using different materials and/or technology?

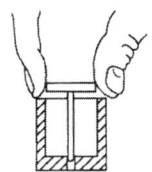
VAT-4: Ease of Assembly

Can even more cost be designed out of this product—and more quality designed in—by making it easier for its parts to be assembled?

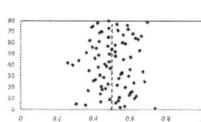
VAT-5: Range

How many values support the same attribute? Can these be reduced?

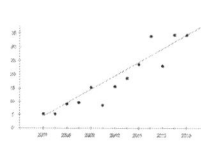
VAT-6: Trend

When attribute values are plotted, can a direction and pattern be observed? Do these observable patterns reveal any additional reduction opportunities related to variety effectiveness?

The fact is, negative variety and its optimal solution are often hidden in the most mundane data—the attributes themselves. Applying the Six VATs provides a process for discovering that. Used in conjunction with the systematic *3-View Analysis* described in the next chapter, these six tools raise and answer many important questions about why and how negative variation occurred in the first place. Resultant insights lead us to develop specific and often creative solutions that prevent the recurrence of unwarranted variety.

CHAPTER 9

The 3-View Analysis

> Which comes first, the product or the customer?
> The answer is: It's a bad question.
>
> — RICHARD SCHONBERGER, *Building a Chain of Customers*

In Stage 1 (Chapters Six and Seven) you set up and trained your VEP Teams, did early promotion, and made your classification system VEP-capable. You also identified your targeted series—where to begin the in-depth VEP analysis to come. In short, you prepared the organization for Stage 2.

Stage 2 began in Chapter Eight when you learned the Six VATs, VEP's analytical tools. These tools are next put to use by three separate but linked teams: market analysis, product structure analysis, and parts type analysis. Known collectively as *The 3-View Analysis*, this process is the diagnostic core of the VEP methodology. The application of the Six VATs can and should extend to control points and processes (Chapter Ten). This chapter, however, focuses exclusively on the first three perspectives (Figure 9.1).

Why Three Views?

The multi-view approach of VEP's *3-View Analysis* is based on the recognition that the triggers or causes of negative variety are multiple—and they are hard to trace. They are more easily discerned through a systematic inquiry process that questions variation in the context of the entire company.

Through the triple lens of this analysis, teams discover redundancies, overlaps, and other inefficiencies in how the company differentiates its products and parts, puts them together, and markets them. Again and again it asks: Is this a customer-driven or an internally-triggered variation?

Figure 9.1. **VEP Methodology: Stage 2/The 3-View Analysis**

STAGE 1 **Plan and Prepare for an Effective VEP Implementation**	STAGE 2 **Identify Reduction Opportunities by Applying the Six VATs**	STAGE 3 Coordinate and Schedule Reduction Proposals	STAGE 4 Implement Improvements and Sustain a VEP Mindset
Step 1 Select a Steering Team that then sets up the other VEP Teams *(Management)*	**Step 1a** Undertake a VEP analysis of market offerings and their characteristics and make reduction proposals *(Market Analysis Team)* **Step 1b** Undertake a VEP analysis of parts as part of the product architecture and make reduction proposals *(Product Structure Analysis Team)* **Step 1c** Undertake a VEP analysis of parts by parts type and make reduction proposals *(Parts Type Analysis Team)*	**Step 1** Coordinate and consolidate reduction proposals *(Steering Team)*	**Step 1** Implement approved reduction proposals *(all VEP Teams)*
Step 2 Conduct VEP training for teams and begin general awareness training *(Education and Methods Team)*	**Step 2*** Undertake a VEP analysis of transactions that support parts, products, and market offerings and make reduction proposals *(Control Points Reduction Team)*	**Step 2** Qualify, approve, and prioritize reduction proposals *(Steering Team)*	**Step 2** Set up a VEP Preventative Monitoring Calendar and continue to educate a VEP mindset *(all VEP Teams)*
Step 3a Find and reduce parts with low-resistance to change *(Early Victories Team)* **Step 3b** Assess, clean up, and upgrade parts classification system *(Parts Type Analysis Team)* **Step 3c** Begin to analyze and revise company policies and practices *(Policy Analysis Team)*	**Step 3**** Undertake a VEP analysis of processes and make reduction proposals *(Process Reduction Team)*	**Step 3** Schedule approved reduction proposals on a VEP Implementation Calendar *(Steering Team)*	
Step 4 Target a priority product as the starting point for Stage 2 reduction analysis *(Steering Team)*			

Stage 2 steps are performed by the 3-View Analysis Teams.

* *The Control Points Reduction Team works independent from other VEP Teams—and can start its work in Stage 1 (see Chapter Ten).*
** *This team is an optional step, dependent on current levels of operational complexity and improvement.*

The 3-View Analysis

While the analysis in this chapter focuses largely on parts, products, and market offerings, the principles are universally applicable, no matter the industry or venue. Bear this in mind, especially if your company profile does not match the one we present. Service offerings, SKU portfolios, IT applications, distributor brand nomenclature—all can benefit from 3-View Analysis. The three-view approach drives the analysis into the telling detail of negative variety, suggests specific ways to rid the system of as much of it as possible, and devises smart, workable guidelines that prevent its recurrence. Along the way, causes of negative and positive variety are tracked and a balance point struck so that *effective variety* is achieved.

Do We Really Need This Difference?

The single driving question in VEP analysis is: Do customers really want or need this difference? Do they really want or need another product series? Another ever-so-slightly-different model? Another subassembly? Another part? Each 3-View Team asks that same question from its own independent framework of inquiry. Their respective answers expose opportunities, great and small, for reduction. Each view provides another piece of the puzzle, another way of understanding how product diversification has, over time, grown into organizational complexity and cost. Broadly speaking, the procedure followed by each team is the same:

- Identify what is the same and what is different.
- Identify the reason for the difference.
- Validate or dispute each of the differences by applying the Six VATs.
- Where the difference is unwarranted, develop a proposal for eliminating it.

Each team develops its own set of recommendations for de-complicating the system, reducing negative variety, and minimizing causes so they will not recur. Each team sends its improvement proposals independently to the Steering Team—which then assesses, coordinates, and integrates all proposals into an overall change schedule (Stage 3). If the proposals can be implemented easily— having low resistance to change—the Steering Team expedites them. If more time, patience or research is needed, the implementation of the improvement gets slated for later in the schedule. In the end, the company gets the best ideas from across all VEP teams.

We will now look at each of the three views teams in detail.

View One: Market Analysis

VEP's market analysis seeks to understand—from the marketing perspective—what makes the company's products different. Which specific customer requirements, niche characteristics, features, elements or functions differentiate one product (or product line) from another?

We address each of these differences by applying VAT-1: Unique vs. Shared and asking the first question: Does the customer really need or want this difference? Is the difference valid? Is this variation customer-driven or internally-triggered?

- If the difference is not a valid customer requirement, we apply the other VATs in the effort to get rid of that difference.
- If the difference is a valid customer requirement, we move to the second VAT-1 question: Is there some less unique or differentiated way to meet this requirement? We apply the other VATs to find out.
- In either case, we apply the other VATs.

At the conclusion of VEP's market analysis, company catalogs may list more products, fewer products—or exactly the same number of products as before VEP. Whichever the case, customer choice is enhanced or even extended. Product distinctions are now validated through analysis; they are clearer, cleaner, and produce more profit. Overlaps are removed, slight or non-required differentiations are blended, and the internal variety needed to support or make these product lines is reduced. Parts count within and across products, as evidenced by the VEP Parts Index, has been reduced.

Canadian-based Venmar Ventilation is a case in point. A market leader in manufacturing built-in residential ventilation/air exchange products, Venmar works with a large chain of retail distributors, under multiple brands. VEP's market-view was the company's first line of analysis attack, due to the growing brand and features confusion in the buying public across the then 149 Venmar products. Even its distributors were challenged to recognize—let alone sell—the differences. Venmar engineers joined with marketing to reorganize the selection, using the Six VATs, starting with: Do we really need this difference? The result was nearly a 40% reduction of product names and the simultaneous consolidation of brand strength and customer choice—not to mention reduced operational complexity.

VEP's market analysis directs us to appreciate and, where possible, improve the way in which product diversification occurs within the marketing function. Companies are constantly adding new products—

sometimes as part of a larger product diversification strategy, sometimes just to test a new market, and sometimes as a startled reaction to an unforeseen competitive threat.

The fact is, marketplace pressures do not always allow a business to develop its product lines in a wholly integrated and rational manner. Product introduction may be haphazard—and when not properly managed, the effects can sink the enterprise. Levels of new products may overlap in capabilities and characteristics with existing products. Dozens of products may linger well beyond their productive life cycles, hanging around in the catalog, the warehouse, and on the shop floor, virtually unnoticed. When sales finally come to a complete halt, they fade but may never be obsoleted.

The goal of VEP's market analysis is two-fold: 1) to identify and remove redundant, declining, obsolete or overlapping products, and 2) to identify and enhance the products that are growing. The process also generates guidelines and practices to ensure that only products that represent the *most effective* variety get introduced in the future.

Your Nomenclature: The Groundwork

You are ready to begin—or almost. Let's check one more thing. Did you standardize the nomenclature related to your product hierarchy? Perhaps you accomplished this in Stage 1, in conjunction with standardizing the terminology in your parts classification system. If not, you need to do it now. What is the name given to the company's most comprehensive level of product? Is it called a *product family* or a *product series*? A product line or a product group? How is the next tier designated? Do some people call it the series level, but others describe it as product types? What about the next layer? Is it referred to as the *model* or the *type* level? And the next?

Neglecting to agree on a common way to describe levels of product significantly impedes the analysis work of this team. A great deal of wasted time results from not rooting out the unwarranted variety in the terms used to refer to products. Let's look at the situation at PUI.

PUI's Product Hierarchy

The VEP analysis process begins with the product line called the *targeted series*, designated in the preparatory steps of Stage 1. At PUI, that product line was Series 7. See Figure 9.2 for the company's Top 20 list.

Figure 9.2. **PUI's Top 20**

PUI'S 20 TOP SERIES			
Series	7*	Series	73
Series	12	Series	17
Series	33	Series	137
Series	1	Series	46
Series	11	Series	28
Series	8	Series	101
Series	4	Series	148
Series	53	Series	79
Series	212	Series	85
Series	97	Series	200

* PUI's Targeted Series

The Market Analysis Team, under the leadership of Willy Sparks, prepared for its analysis tasks by first clarifying PUI's product levels into the product hierarchy and standardized the nomenclature (Figure 9.3).

Figure 9.3. **PUI's Product Hierarchy: Nomenclature of Levels**

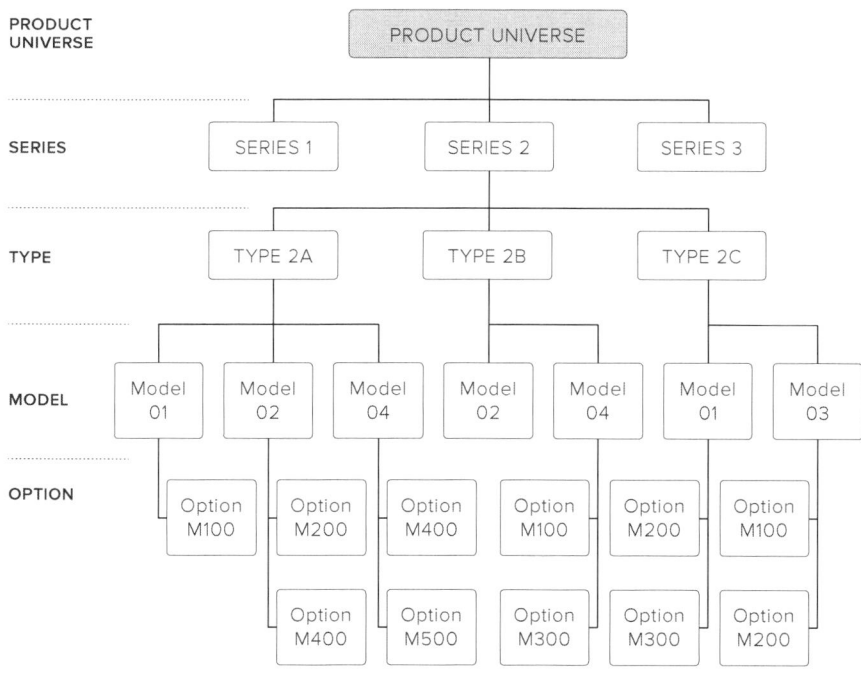

Series. At the highest level, like-products at PUI are gathered into product groups called *Series*. These groups carry the same number designation, such as *Series 4* or *Series 7*. PUI's product series includes: explosion-proof switches, NEMA-4 controls, recorders, and heat-trace controllers. At the end of last year, 50 different product series were considered active at PUI.

Type. Within a given product series, PUI products with similar properties are considered the same *product type* and are designated by a letter. For example, products with remote temperature sensing properties carry an A designation. *Type B* products have local temperature sensing properties, and *Type C* products control pressure. Therefore, a product designated "Type 8A" is part of Series 8 and has a remote temperature sensing capability.

Model. Products at PUI with the same sensing ranges are considered the same model and are designated by a two-digit code. For example, all Model 01s have a sensing range of 20 to 200 PSI (pounds per square inch), whereas all Model 03s have a 50 to 650 degree Fahrenheit range. Therefore, the same model can appear on one or more *Types* within one or more Series. For example:

Product 7A-01 = Series 7, Type A, Model 01

Product 4A-01 = Series 4, Type A, Model 01

Product 7C-01 = Series 7, Type C, Model 01

Options. At PUI, almost any type and/or model can be ordered with one or more options. Options include special enclosure coatings, optional sensor materials, custom labels, etc.

Calculating the Combination Magnitude

With the terminology clarified and agreed upon, PUI's Market Analysis Team was ready to address the magnitude issue. Sparks asked team members to choose any three series in PUI's Top 20 list and calculate the number of products a customer *could* order, based on the types, models, and options offered in those three series. They were amazed at the results: A vast number of products that customers could buy—in some cases, over 50,000—have only the slightest distinctions between them. (See Figure 9.4 for one such set of calculations.) The team realized that, if this applies to only three series, the total number of possible combinations based on PUI's complete line of products—all 50 series—would be astronomical. They got fired up and eager to start.

Figure 9.4. **Example: Combination Magnitude on Three of PUI's Top 20 Product Series**

SERIES	NUMBER/ TYPES	NUMBER/ MODELS	NUMBER/ OPTIONS	TOTAL POSSIBLE COMBINATIONS
11	22	38	22	18,392
1	24	46	24	26,496
8	9	37	17	56,661

Total Possible Combinations Across These Three Series = 50,549

Comparing Product Attributes

The next step in market analysis is to develop a product-attribute matrix that identifies the main market-driven/customer-driven characteristics of each product in your product universe. Begin with your targeted series.

When, for example, the PUI team created a Product Attribute Matrix for Series 7 (targeted series), they identified the following attributes as relevant: enclosure ratings, electrical ratings, the number of control values, and agency approvals. They plugged these attributes into the matrix for Series 7 and then added other product series that shared the same primary properties. The team went a step further and listed these products by their respective percentage of sales in descending order (Figure 9.5).

Team members were now able to make comparisons and search for reduction opportunities, using the Six VATs. They applied VAT-1: Unique vs. Shared to systematically question the overlaps, redundancies, and slight variations they found in many primary characteristics. Many chances to standardize across the various series became apparent. Using VAT-3: Multi-functionality & Synthesis, they brainstormed ways to streamline certain series or collapse one series into another. They were surprised at how easy the matrix made these applications. Its format facilitated discussion and kept brainstorming focused as they considered ways to simplify or eliminate product segments without negatively impacting customer selection.

When one matrix was completed, the team ratcheted down to the next level in the hierarchy (product type), developing a new matrix for analysis. After that, the focus shifted down to the model level. The search continued for further areas of product consolidation or differentiation that did not diminish customer selection.

Figure 9.5. Market Attribute Matrix (Partial): Series Level

SERIES DESIG-NATION	% OF SALES	MARKET ENTRY DATE	AGENCY APPROVALS	ENCLOSURE RATINGS	ELECTRICAL RATINGS	NUMBER OF CONTROL VALUES
7	12.3	8-83	UL, CSA	NEMA 4	15A- 125/250/480 VAC Res	1 or 2
12	11.9	11-01	UL, CSA, FM, ISSEP, BASEEFA, NACE	NEMA 4, 7, 9	15A- 125/250/480 VAC Res	1 or 2
33	9.0	6-54	UL, CSA	NEMA 4, 13	15A- 125/250/480 VAC Res	1, 2 or 3
4	5.5	3-77	UL Recognition	None	15A- 125/250 VAC Res	1
53	3.9	10-01	UL, CSA Optional	NEMA 4	15A- 125/250/480 VAC Res	1 or 2
1	6.5	6-09	UL, CSA Optional	NEMA 1, NEMA 4, 7, 9 Optional	15A- 125/250/480 VAC Res	1 or 2
73	1.7	10-04	UL Recognition, CSA	None	15A- 125/250 VAC Res	0, 1 or 2
8	7.0	5-72	UL, CSA	NEMA 4, 4X	15A- 125/250/480 VAC Res	1
97	2.6	9-89	UL (most models), CSA Optional	NEMA 4	15A- 125/250/480 VAC Res	1

LEGEND: DEFINITIONS OF ACRONYMS IN ABOVE MATRIX

A. **Approval Agencies**
  1. BASEEFA = British Approval Service for Electrical Equipment and Flammable Atmosphere
  2. CSA = Canadian Standards Association
  3. OPTIONAL = An approval rating is not standard on this product; it may be purchased as an option.
  4. FM = Factory Mutual
  5. ISSEP = Institut Scientifique de Service Public
  6. NACE = National Association of Corrosion Engineers
  7. UL = Underwriters Laboratories
  8. UL APPROVED = Higher rating than UL Recognized
  9. UL RECOGNIZED = Lower rating than UL Approved

B. **Enclosure Ratings**
  1. NEMA = National Electrical Manufacturers Association
  2. NEMA 4 = Weatherproof
  3. NEMA 4X = Waterproof
  4. NEMA 7 = Explosion Proof
  5. NEMA 9 = Explosion Proof
  6. NEMA 13 = Explosion Proof

C. **Electrical Ratings**
  1. 15A = 15 Amps
  2. VAC RES = Volts Alternating Current Resistive: 25, 250, or 480

D. **Number of Control Values**
  (Number of values switch can control at once)
  1. Two control values = switch turn-off at 15E and 45E
  2. 0 control value = value indicated but no control over it

Along the way, the PUI team regularly updated the VEP Parts Index (Chapter Three) that had been developed at the outset of the analysis to track each proposal's impact on parts count reduction. They were also careful to review each improvement idea with the appropriate stakeholders as part of the validation process. After several rounds, the team submitted its first-pass reduction proposals to the Steering Team (Figure 9.6).

Figure 9.6. **PUI Market Analysis Team: Reduction Recommendations (First Pass)**

RECOMMENDATIONS	IMPACT ON PARTS NUMBER COUNT	IMPACT ON INDEX	IMPACT ON MARKET
1. Eliminate eight different product series: Series 76 Series 16 Series 68 Series 12 Series 53 Series 5 Series 31 Series 2	Eliminate 350 part numbers	Reduce index value by 3182 (43 parts types X 74 part numbers in the models eliminated when series are eliminated)	No discernible impact on market. PUI retains ability to meet all existing market requirements.
2. Eliminate five product types in remaining series: Type L Type S Type P Type V Type Q	Eliminate 65 part numbers	Reduce index value by 455 (13 parts types X 35 part numbers across product types recommended for reduction)	The same or nearly the same functional capabilities are found in other PUI products.
3. Eliminate 16 individual models across remaining series: Model 701 Model 153 Model 702 Model 152 Model 703 Model 151 Model 704 Model 93 Model 190 Model 92 Model 168 Model 77 Model 167 Model 73 Model 154 Model 71	Eliminate 90 part numbers	Reduce index value by 792 (24 parts types X 33 part numbers across the models recommended for reduction)	Marketing says no ramifications; changes are "invisible" to PUI customers.
Total Part Numbers Eliminated: **505**	**Total Reduction of VEP Parts Index:** **4,429**	**Total Market Impact:** PUI customer selection remains constant at lower cost and less systems complexity	

Implications: Market Analysis Process

For companies in the habit of adding products in a less than rational manner, VEP market analysis is an illuminating process. It exposes the tangible consequences of proliferating products in the absence of an official proliferation strategy. In the process of finding reference points and probing for differentiation, the team gains a cross-product perspective that refutes or validates, unequivocally, the answer to the driving VEP question: "Does the customer really want or need this difference?"

Without analysis, the result is often an ever-lengthening roster of products on series, type, and model levels that represents only negative variety—unwarranted, costly, and complex. As a result of analysis, team members recommend combining (or loping off) certain products—and they do so confidently, guided by this structured, systematic process. They are assured that customer selection will not be jeopardized.

Spreading Out the Reduction Net

There is another important implication of the market analysis process, one that holds true for all three 3-View efforts. Looking at the proposals that PUI's Market Analysis Team submitted (Figure 9.4), you might think you see an inconsistency—several of the product series recommended for elimination had not appeared on PUI's Top 20 list (Figure 9.2). Notice that Series 2, 5, 16, 31, 68, and 76 were not in the Top 20—but do appear as reduction recommendations.

This is not an anomaly. The process is designed to magnetize or link other items that have characteristics in common with those of the targeted series. All VEP's analytical procedures do this—exactly what we see in the Market Attribute Matrix (Figure 9.5). Anchored in the primary characteristics of the targeted series (Series 7), the matrix mechanism pulled in all linked series—some of which were not in the Top 20. As the search for overlaps and redundancies extended into type and model levels (still anchored in Series 7), other non-Top 20 products were snared. In this way, the analysis touches most, if not all, of the products in the company's product universe. The analysis net spreads out, capturing instances of negative variety throughout the company and holding them up to systematic scrutiny.

Now let's see the analysis process through the lens of the next 3-View team: Product Structure.

View Two: Product Structure Analysis

The purpose of Product Structure Analysis is to understand and improve how product function requirements are fulfilled through diverse parts and components so that a viable, integrated, and profitable unit called a *product* results. Similar to the market view, this analysis approach begins by identifying how products are currently configured—and ends by recommending changes that streamline and strengthen this.

Using the product Bill of Materials (BOM) as a base, the examination takes place on the model level, focusing on how parts mate and components are configured—and on the structural or geometric level for differing parts values. In this way, the analysis validates or refutes the architectural logic by which a company's products are developed, both as a population and in particular. The result, after an assiduous application of the Six VATs, is a line of products that is more variety-effective and less costly.

A product, as VEP defines it, is: *a group of elements or parts that, when fabricated and assembled, fulfill specific requirements.* Product structure analysis focuses on the *relationship* between a product's constituent elements. That relationship is the chief determinant of negative or positive variety. This refers, variously, to: the product structure, product geometry or product architecture—the way parts are used to form the product. The goal is structural simplification, with equal or enhanced functionality, using fewer parts.

The process of structural analysis begins on the model level: how product functions are fulfilled through combinations of parts, components, and subassemblies. In a manner similar to market analysis, the inquiry spirals out, steadily and systematically—and, through its iterations, eventually encompasses virtually every company model.

Key Elements of the Product Analysis Procedure

- The team starts with the targeted series—the jumping-off point for all inquiry—stepping through the following procedure:
- Choose a priority model in the targeted series, based on an agreed-upon set of selection criteria—highest sales volume, volume shipped, and/or product complexity (as a function of the number of parts).
- Identify where that same model is used throughout the company's product universe.

- Generate a BOM for each model.
- Calculate a VEP Parts Index across these models.
- Physically gather the constituent BOM parts for each model.
- Create a physical visual layout of these BOMs, laying out the constituent parts, side by side, model by model, creating an actual "exploded" view of the products (Figure 9.7). Facilitate this review by shrink wrapping the parts display onto cardboard (this also makes it easy to carry to and from meetings).

Figure 9.7. **Visual Layout of BOM**

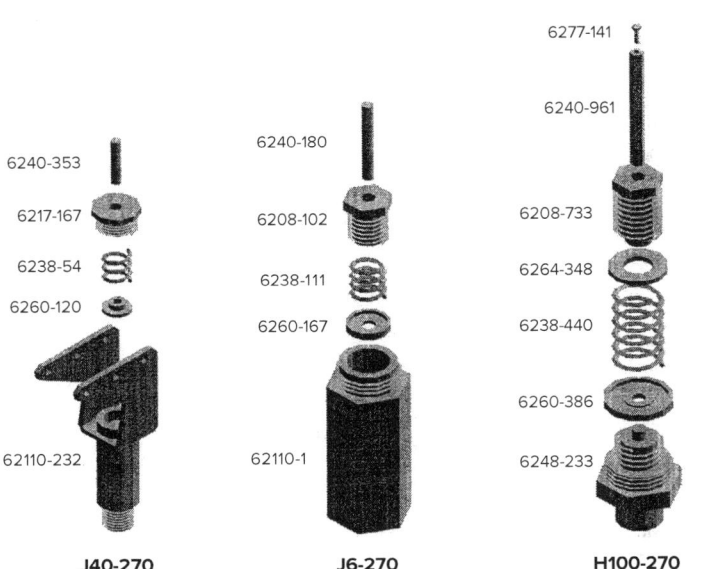

Guided by the Six VATs and their associated questions, the team focuses on the physical, visual, *real-time* examination of these parts, laid out in the above manner. Team members now look, they compare—*and they think!* As the parts are visually compared across models, important questions surface, as do improvement proposals.

Let's watch the process as it unfolded with the PUI Product Structure Analysis Team and its focus on five 02 Models, a popular temperature control product. The team generated BOMs on all 02 Models. See Figure 9.8 for a partial BOM and Figure 9.9 for the VEP Parts Index (1,679) based on that partial BOM. Ultimately, when they calculated all BOMs across the complete range of 02 Models in all related Series, the Total Parts Index came to 93,445 total occurrences.

Figure 9.8. **Partial BOM: One of the 02 Models**

PART NUMBER	DESCRIPTION
013-65	Screw 6-32 X 1/4 B/H S/S
017-69	Screw 6-32 X 7/8 P/H Sems S/S
017-85	Screw 8-32 X 1/4 R/H S/S
52-123	Overtravel Housing
6154-1	Cover
620-25	Overtravel Button
625-24	Insulator Switch
628-56	Adjusting Screw
611-55	Computer Nameplate
621-54	Switch Bracket Assembly
627-38	Plunger Stop
623-29	Bellows Bracket Mating
628-19	Spring
638-29	Spring
628-27	Bellows 1 1/2 Heavy Press
626-16	Spring Guide

Team members then applied the Six VATs. VAT-1: Unique vs. Shared revealed that four of five pressure connections were *nearly* identical. Probing deeper, they discovered that, in three cases, there was no valid customer-driven reason for the difference. Commonization became an option.

VAT-2: Modularity also produced strong results. Looking for options in organizing standardized parts into interchangeable units, the team discovered that standardizing the pressure connection across three models opened the door to a modularized pressure assembly in all of them. Doing so would also obviate the differences between Series 8 and Series 11—as well as entirely eliminate the need for one of these series. The idea was tagged. John Mandella volunteered to ask the Market Analysis Team if this reduction proposal would create any customer problems.

In another instance, the product structure team saw that an identical interchangeable unit could be shared across Series 17 and Series 28, if certain parts in both series were comparably standardized. That idea also got tagged. Tricia Moodley checked in with various stakeholders for validation.

Figure 9.9. VEP Parts Index: Partial BOM for Models 02 in Series 7, 8, 11, 33, and 97

PARTS TYPE	PART NUMBER	MODEL 7-02	MODEL 8-02	MODEL 11-02	MODEL 33-02	MODEL 97-02	TOTAL PART TYPE OCCURRENCES
Screw	013-65	2	2				
	017-69			2	2		3
	018-85					2	
Overtravel Housing	52-108		1		1	1	2
	52-123	1		1			
Cover	6154-1	1	1	1	1	1	1
Overtravel Button	620-25	1	1	1	1	1	1
Insulator Switch	625-24	1					2
	625-42		1	1	1	1	
Adjusting Screw	628-56	1	1	1	1	1	1
Computer Nameplate	611-55	1	1	1	1	1	1
Switch Bracket Assembly	621-67		1		1		3
	621-54	1				1	
	621-90			1			
Plunger Stop	627-38	1	1	1	1	1	1
Bellows Bracket Mating	623-29	1					2
	623-27		1	1	1	1	
Spring	628-19	1	1	1		1	2
	628-29	1		1	1	1	
Bellows	628-27	1	1				3
	628-28			1	1		
	628-31					1	
Spring Guide	626-16	1	1	1	1	1	1
							23
Total Parts Count		15	14	15	14	15	73

VEP PARTS INDEX (Partial BOM) = 1,679 (73 × 23)

The team also found a significant application of VAT-3: Multi-functionality & Synthesis. For 15 minutes of stark silence, team members stared at the shrink-wrapped layout board. Suddenly Gerry and Danielle leapt to their feet, pointing to the spring guide. They both saw how the functions of the spring guide and plunger for *all five* models could get collapsed into a single part—as shown in Figure 8.5 in the previous chapter. The parts count impact for PUI was huge because spring guides and plungers were used in practically every PUI model. This idea was marked for review by the Parts Type Team.

Figure 9.10. **Model 02 Reduction Recommendations (Partial List)**

PRODUCT STRUCTURE REDUCTION RECOMMENDATION	PARTS IMPACT	SUPPORTING DATA & REMARKS
1. Standardize on pressure connection 5832-141 and eliminate pressure connection 5832-120	Would eliminate a total of four part numbers: 5832-120 assembly and 3 part numbers that make it up (5832-102, 6471-102, 6471-105)	This change would not affect model performance or specifications.
2. Standardize on higher barrier switch: 5269-618	Would eliminate 2 insulator part numbers (6107-921 and 6107-922)	This change would increase overall switch height by 1/16". UL, CSA, and marketing have already approved.
3. Purchase housing with brackets already in place	Would eliminate 1 set of bracket (443-0-152) plus related spot welding processes	Supplier would perform spot weld (piece price goes up $.04) or re-tool to make bracket part of housing (one time charge of $2,500). Marketing sees no negative effect.
4. Add optional mounting hole locations on switch mounting bracket 8847-09	Would eliminate 5 bracket part numbers with non-standard mounting hole locations and all optional drilling	Supplier says could easily make this change. Cost per piece increase would be $.07. Marketing says *no problema*.
IMPACT ON PARTS COUNT AND MARKET		
Total Part Numbers Eliminated: **13**	Total VEP Parts Index Reduction: **954**	Total Reductions in Production Processes: **12**	Total Market Impact: No Negative Impact

As the team completed its first round of analysis, the VEP Parts Index had already shifted down by 12%—from 93,445 to 82,232. As with all VEP improvement ideas, each proposal was systematically validated or invalidated against stakeholder concerns. If the idea turned out to be problematic, it was held in reserve. Validated ideas went to the Steering Team for review and coordination. See Figure 9.10 for a partial list of PUI's Product Structure Team's reduction recommendations for Model 02—and the impact of those recommendations.

Implications: Product Structure Analysis

VEP's product structure analysis obliges us to scrutinize and compare the form, fit, and function of all the parts in a given model—within and across product families. This simple but structured approach allows teams to brainstorm possibilities for eliminating negative variety, even as they are guided by a systematic application of the Six VATs. By the time this analysis is completed, each part—and its relationship to all other parts—has been studied, probed, and understood in terms of the functionality of both the specific model and its sister models.

This in-depth scrutiny directs a steady, uncompromising light onto the company product-differentiating tendencies and biases. It can lead to significant product simplification and a reduction in the number of differing structures across the company's product universe. The technique's extensive use of visual layouts (an exploded view of the physical BOM on a model level) offers a unique window for engineers to reconsider and reconstruct the decision-making process by which a product was designed. Doing this can trigger valuable insights into the practices that do—and don't—favor positive variety. In the process, teams identify internal and external triggers of product structure complexity and find ways to reduce or eliminate these causes.

We will now move on to the final of the three analytical views—Parts Type.

View Three: Parts Type Analysis

The objective of the Parts Type approach, VEP's third product-based analytical perspective, is to minimize the total number of different individual parts in company products, *across the board*.

While product structure analysis concentrates on the relationship between a product's diverse parts, parts type analysis looks at *each* constituent part of a product—separately and in near isolation from other parts within that product. It looks at the part as a member of a group of similar but not identical parts and the extent to which attribute values vary within each *parts type* or *parts commodity*. Examples of common parts types include fasteners, housings, trim, handles, labels, knobs, hinges, lids, and brackets.

Parts Type Analysis is an aggressive examination driven by a single question: Do we need another part number? If there are 6,000 different fasteners, the question gets asked 6,000 times. If 50% of the housings have built-in brackets, 30% have no brackets, and the remaining 20% use twelve different kinds of brackets, we want to find out *why*.

If reasons can be found and are validated by stakeholders (customers, engineers, approval agencies, etc.), the variations stay. But if there are no good reasons, they go. The same goes for those seventeen different door hinges in use. The Parts Type Analysis process asks and answers these "why" questions. At its conclusion, many part numbers are eliminated and a wider appreciation of parts type economies is understood.

For VEP purposes, a parts type group is defined as: *two or more parts that serve identical function, differing only in the specifications used to fulfill that function*. As a manufacturer of switches and controls, PUI's common parts types include screws, housings, brackets, springs, spring guides, O-rings, lead wires, plungers, fasteners, labels, diaphragms, and bellows—to name a few. At automotive companies, by contrast, the parts types list would include anything from trim, knobs, handles, hinges, mirrors, meters, visors, and windshield wipers to wheels, wheel rims, seats, engine blocks, radiator caps, and, of course, steering wheels. Computer hardware manufacturers have an ongoing need for resistors, transducers, capacitors, printed circuit boards, side panels, front panels, rollers, bases, housings, and so on and so forth—parts types all.

Parts type analysis focuses on the quantity of different part numbers within each parts type. Often the specifications that define parts within a parts type are identical in all but one attribute. It is also not uncommon for that attribute to fluctuate widely in value within that parts type. Furthermore, it is not unusual for no one to know why. Similar functions but dissimilar specifications. Why? Once again, VEP queries: Does the customer really need this difference? Is yet another part number really necessary?

Figure 9.11 shows a work form that one Parts Type Team used early in its analysis to identify an initial part profile. Notice that it requires that we categorize a targeted part as *variable, quasi-variable* or *fixed*. Also imbedded in this work form is a high/low matrix for gauging the relationship between the targeted part's volume of use and level of variation. By applying a nice twist on the Five Whys tool from lean, this initial profile is further strengthened by requiring us to question why a targeted part should—or should not—be allowed to remain in the company's parts inventory. This is a sturdy form that is well worth your study not just for VEP parts type analysis but for all VEP analysis.

Figure 9.11. Sample: VEP Parts Profile Work Sheet

VEP PARTS PROFILE WORK SHEET

Parts Type _____ Your Team _____
Part Number _____ Your Name _____
Class Code _____ Today's Date _____

1. **This part is** *(check as applies):* ☐ Fixed ☐ Quasi-Variable ☑ Variable
2. **In which product type(s) is this part used?** Name all relevant part types by type number or attach a computer printout.

3. **Categorize the part on the matrix.**

HOW IS IT MADE?	WHO SUPPLIES IT? (SOURCE)	WHO USES IT?	WHAT IS ITS FUNCTION? (HOW IS THE PART USED?)	WHAT IS ITS VOLUME OF USE?	WHAT IS THE COST OF THIS PART?
Identify the process:	EXTERNAL Identify:	EXTERNAL ☐ Customer ☐ End User	☐ Instructional (adjustment) ☐ Safety (agency)		
	INTERNAL Identify:	INTERNAL Identify:			

4. **Plot the variety/volume relationship on the Matrix:**

(Volume: Low to High vertical axis; Variety: Low to High horizontal axis; 2×2 matrix)

5. **Do we really need this part?** Ask "WHY?" five times for YES and MAYBE.

☐ YES	Why?	If YES, it's time to seek to reduce. Combine, modularize, etc. It's time for next VEP step.
☐ NO	Explain:	If NO, apply obsolescent procedures.
☐ MAYBE	Why?	If MAYBE, gather the information you need to answer the question with a clear "YES" or "NO" response.

Parts Type Analysis at PUI

Parts Classification Is Task One. The first task of PUI's Parts Type Analysis Team, undertaken at the start of Stage 1, was to clean up the company's existing parts classification system (Chapter Seven).

The Parts Type Analysis Team had come to understand why a messy parts classification system is incapable of supporting the data queries and decision making central to the VEP method. Starting with the parts types in the BOMs of the targeted series (Series 7), the team busied itself removing redundant and obsolete part numbers, creating attribute templates, and inputting specification data into those templates. They continued doing this until the system was "smart" enough (VEP-capable) to support the first cycle of parts type evaluation. Although the other two teams involved in the 3-View Analysis had already begun to tackle their tasks, the work of these two teams was greatly accelerated when the company's classification system reached this level of capability.

Next: The Core Task—Apply the VATs. With an upgraded classification system up and running, PUI's Parts Type Team was ready to tackle its central mandate: parts type analysis and reduction. As its first step, the team generated a Parts Type list (Figure 9.12) and VEP Parts Index on Series 7 so it could identify the parts type with the highest number of occurrences. That is where the analysis would begin. That parts type in Series 7 turned out to be springs.

Figure 9.12. **Parts Types in Series 7 (Partial List)**

PART TYPES IN SERIES 7 (PARTIAL LIST)	
Bellows	Micro-Switches
Brackets	Nameplates
Capacitors	O-Rings
Contacts	Plunger
Covers	Relays
Dials	Screws
Diaphragms	Seals
Gaskets	Spiral Pins
Housings	Spring Guides
Insulators	Springs
Knobs	Switches
Lead Wires	

Because parts type analysis, by definition, touches the company's entire parts type universe, the team turned for help and insight to the tool that gave the widest window on the variation within parts types—VAT-5: Range. In keeping with that technique, team members queried the specifications of each spring attribute, as listed in the spring attribute template they had previously developed—spring material, rate force, outer diameter (O.D.), inner diameter (I.D.), free length, solid length, and material. The specification for each attribute was then formatted into a range dispersion (Figure 9.13, a repeat of Figure 8.12/previous chapter).

Figure 9.13. **VAT-5/Range: Springs Re-Visited by the Parts Type Team**

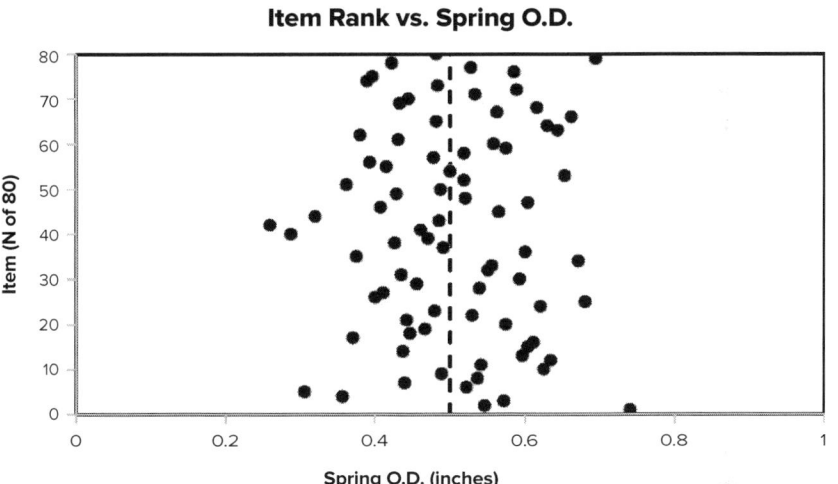

Right off the bat, team members saw the wide range of O.D. values (heavy at the two tails), the gaps in the middle range, and the oddball single-occurrence values scattered throughout. They began to wonder why so many variations in O.D. were necessary. Applying VAT-1 (Unique vs. Shared), the team questioned the validity of each point of variation along the range spread. They asked: Is this specific variation needed? If not, how might it be removed? In some cases, the team proposed reducing the range of variation by making the O.D. value more robust. In cases where the geometry of the product did not allow for that, they proposed standardizing certain values. It soon became clear that many spring part numbers could be shared within Series 7 and also across product lines. With each proposal, the team updated the VEP Parts Index and saw for themselves the potential impact of their ideas on parts count.

An application of VAT-2 (Modularity) helped the team realize that some sub-assemblies in Series 7 could be treated as a parts type. They extended the same concept to other product lines. VAT-3 (Multi-functionality & Synthesis) led the team to consider collapsing two—even three—parts types into one. In fact, the Parts Type Analysis Team came up with the exact same proposal that the Product Structure Analysis Team submitted: Consolidate the spring guide and plunger functions into a single part. See again Figure 8.5.

Figure 9.14. **Parts Type Reduction Recommendations: Screws and Brackets**

RECOMMENDATIONS	PARTS IMPACT	SUPPORTING DATA AND REMARKS
Screws		
1. Eliminate all brass screws; use only stainless steel.	Total number of screws reduced by 9	Actual cost differential is minimal, given that the future volume of stainless steel screws will bring down piece price to within $.01 of current brass screw price.
2. Eliminate all hex head screws; use only flathead.	Eliminates 8 more screws	Design Engineering says no effect. Marketing confirms that customers are willing to accept change. Manufacturing Engineering says it's acceptable to operators plus reduces number of tools required and number of parts locations.
3. Substitute 1/4" screw for all 1/8" screws.	Eliminates 3 more screws	Marketing says no market ramifications. Assembly says 1/4" screws are easier to handle.
Switch Brackets		
1. Eliminate pivot parts 6332-131 and 6332-132. Standardize on using 6332-133 only.	Reduces 2 part numbers	Engineering says 6332-133 will work on all switch brackets.
IMPACT ON PARTS COUNT AND MARKET		
Total Part Numbers Eliminated **22**	Total Index Reduction **3,742**	Total Market Impact: PUI customer selection remains constant at lower cost and less systems complexity.

All the VATs worked beautifully for the team—but VAT-6 (Trend) produced a special kind of magic. Using it, the team traced the growth and overall direction of the specifications of a particular attribute. This is done by associating one attribute with another attribute—or against a time axis. One trend query demonstrated the company's penchant for purchasing stainless steel brackets in lieu of brass in newer products—while another showed that bracket tolerances had been progressively widening over the last three years. Yet another run revealed that part specifications on original equipment manufacturer (OEM) housings had remained unchanged over the past five years in most series but were fluctuating widely in PUI's fiber optic products. Similar to its applications in marketing, the trend technique exposed otherwise hidden, negative tendencies and biases.

Nearly every session of the Parts Type Analysis Team produced a bumper crop of reduction ideas. Invariably, a good percentage of them did not pass muster when checked out with stakeholders. But enough did. Figure 9.14 shows a set of proposals related to screws and brackets that made it to the Steering Team review. The team also developed inventive proposals for preventing future proliferation. One of these, based on an article a team member read about Nissan, was to set up "border guards," ex-VEP team members charged with blocking incursions (new parts) that would needlessly elevate the company's parts count.

Figure 9.15 shows a meeting template one VEP company required its VEP Teams to use as a guide and support. In this sample, the Market Analysis Team is just finding its way; as this verbatim language demonstrates, they have already absorbed much of the VEP nomenclature and focus and are off to a strong and systematic start. Take a look and consider using a template like this for all your analysis teams.

Implications: Parts Type Analysis

We return to the automotive industry and its potential for mushrooming parts inventory. The following variations in parts types were tracked across a vehicle pool of 100 models:

- 1,200 different kinds of floor carpets
- 437 different kinds of dashboard meters
- 300 different kinds of ashtrays
- 110 different kinds of radiators

Figure 9.15. **VEP Team Meeting Form: Market Analysis**

VEP TEAM MEETING FORM

PAGE 1

Team: *Market Analysis*
Team Leader: *Dawn Houston*
Today's Date: *08/03* **Day:** *Tuesdays* **Time:** *9:00-10:00*
Team Members: *L. Tolemei, H. McNeeley; J. Martinez; D. Higgins*
Room: *M2*
Notes by: *H. McNeeley*

A. Team Objective:

To evaluate PUI's product universe from a market prospective in order to identify reduction opportunities in the existing product lines that do not endanger the company's market position or limit customer selection.
- Identify & record any methods and tools your team utilized or developed.
- Make sure we specify how we as a team reached the decisions we did. Share charts and forms.

Guidelines for Every Meeting:
1. Set agenda in advance and follow it
2. Stay open to new ideas
3. Set incremental targets
4. Complete this sheet and submit a copy to VEP Steering Team Lead.

Distribution to Others:
- ☐ G. Elliott
- ☐ M. Davies
- ☐ J. Henriquez
- ☐ All Team Members

B. Objectives for Today's Meeting	Time Allotted
1. Develop first draft of series reduction list (Helen).	20 minutes
2. Identify impact of reduction list on revenues (Jose).	20 minutes
3. Scope out work plans for next 3 weeks (Dawn).	10 minutes

C. Today's Actions and Questions	D. Today's Outcomes and Decisions
1. Reviewed series worksheets and brainstormed ideas for reduction of # of series. 2. Created list from #1 that identifies impact on revenue, inventory, etc. 3. Outlined focus of meetings for next three weeks and, in general, the time frame for completing market analysis teams work. **Questions** 1. What is our Product Obsolescence policy? 2. Do we share it with customers? How?	1. Developed first pass list of reduction candidates. Agreed to bring pros and cons for eliminating each on post-its to next meeting for discussion. 2. S. Miller will also get # of dedicated parts for each series in list. 3. Agreed to use this form to keep track of ideas, future recommendations for series, models, and types. 4. Assigned team members various responsibilities for keeping team on track for next three weeks. • J. Martinez- Complete type matrices for series 02 by 8/16. • L. Tolemei- Obtain inventory value as of 08/03 for all series & bring to next meeting on 08/10. • H. McNeeley- First draft of comparison criteria for models by 08/23. • D. Higgins- Complete series form by 08/30

VEP TEAM MEETING FORM

PAGE 2

Team: *Market Analysis*

E. Methods, Tools, Remarks: Identify any charts or forms you developed or used during today's meeting.

We made a list of series to consider for reduction (1st draft).
We also created a form that lists series, last year's revenues for series, and current inventory dollars in hand.

F. Policy Remarks: Identify any policy that needs to be clarified or created from today's meeting.

We need more guidance on life cycle decisions. When is a series ripe for reduction? What are some general guidelines we could apply to determine when a series should be eliminated?

G. What control points, parts or processes did you eliminate today?

None. We are still in preliminary stages of evaluating.

Agenda Items For Our Next Meeting	Time Allotted
1. Lorraine reports on value for all series. (L: Please provide handouts of your analysis).	15 minutes
2. Jose will show us a sample Type Matrix to critique before he reports on 8/16	10 minutes
3. Helen leads us in brainstorming criteria for models	20 minutes
4.	
5.	

One midsize sedan alone had 62 different kinds of electrical harnesses associated with it, representing in their totality 17 yards of harness storage bins.

Legitimately, we ask: Does this level of variation create and sustain markets—or suffocate them? Will the customer buy that difference? One automaker got part of its answer when it discovered that 50% of the variations that resulted in its prodigious 2,200-model range contributed only 5% to total sales.

Systemic Problem—Systematic Approach

While VEP's 3-View Analysis may not be able to ascertain the exact cause of variations (numerous factors can contribute), it can vigorously query why these variations must remain and help root them out. It shows you precisely where negative variety resides in your company.

Runaway proliferation places huge burdens on the organization. Nowhere is this better appreciated than in calculating variations within each parts type. But it is not just hiding in your parts type pool. It resides in a vast array of front-end and in-process decisions that cause another new product line, another new product, and another new part to enter the system. Variation and the complexity that follows in its wake are *embedded* in all company systems. Because it is everywhere—because the problem is systemic—it can only be successfully attacked through a systematic approach.

The three-prong offensive of VEP's 3-View Analysis brings into balance the three competing perspectives of product marketing, product design and development, and parts type proliferating.

There is another major outcome from this analysis process: Company teams learn to work strongly in their independent roles even as they focus on the larger goal of de-complicating the organization. If you are already accustomed to working in teams, you will move quickly to the task. If you and your colleagues are new to team structure, the VEP Team architecture provides a structured way to get started, with clear goals, objectives, and tools.

In the next chapter, we round out Stage 2 by applying the Six VATs to processes and control points—an option appropriate to most companies.

CHAPTER 10

Reducing Downstream Complexity
Unwarranted Processes and Control Points

Froggie didn't notice the water he was sitting in begin to boil. It had been such a nice warm pool for so long.

This chapter brings us to the final segment of Stage 2 of the VEP Methodology. To this point, you have sought out opportunities to reduce negative variety in market offerings, product structure, and the parts universe by applying the Six VATs. Now it is time to look at operational processes and control points—the layers of downstream complexity triggered by the mere existence of markets, products, and parts. (See shaded area in Figure 10.1.)

The discussion begins with the introduction of terms. VEP defines a process as: *a sequence of events, steps or activities performed in order to reach a specific outcome.* They always reflect F-Costs (Function Costs) as discussed in Chapter Three. For example, on the production floor in a machining company, appropriate F-Cost processes would include rough and final machining, heat treat, deburring, final machining, and so forth. Negative variety in processes refers to those that are redundant or simply not needed—not needed because the product (or service) configuration has changed and with it the associated parts. Therefore, when we remove a market offering, product or part as the result of our 3-View Analysis, the associated processes are eliminated as well.

Similarly, *control points are the activities that support a product, part, service or process.* In the same machining company that would include: purchasing; invoicing; scheduling; incoming, in-process, and final inspection; supplier certification; moving, storing, and retrieving parts; and so on. Control points add cost but no direct value to the customer.

Figure 10.1. **VEP Methodology: Stage 2/ Reduction Analysis for Processes and Control Points**

STAGE 1 Plan and Prepare for an Effective VEP Implementation	STAGE 2 Identify Reduction Opportunities by Applying the Six VATs	STAGE 3 Coordinate and Schedule Reduction Proposals	STAGE 4 Implement Improvements and Sustain a VEP Mindset
Step 1 Select a Steering Team that then sets up the other VEP Teams *(Management)*	**Step 1a** Undertake a VEP analysis of market offerings and their characteristics and make reduction proposals *(Market Analysis Team)* **Step 1b** Undertake a VEP analysis of parts as part of the product architecture and make reduction proposals *(Product Structure Analysis Team)* **Step 1c** Undertake a VEP analysis of parts by parts type and make reduction proposals *(Parts Type Analysis Team)*	**Step 1** Coordinate and consolidate reduction proposals *(Steering Team)*	**Step 1** Implement approved reduction proposals *(all VEP Teams)*
Step 2 Conduct VEP training for teams and begin general awareness training *(Education and Methods Team)*	**Step 2*** Undertake a VEP analysis of transactions that support parts, products, and market offerings and make reduction proposals *(Control Points Reduction Team)*	**Step 2** Qualify, approve, and prioritize reduction proposals *(Steering Team)*	**Step 2** Set up a VEP Preventative Monitoring Calendar and continue to educate a VEP mindset *(all VEP Teams)*
Step 3a Find and reduce parts with low-resistance to change *(Early Victories Team)* **Step 3b** Assess, clean up, and upgrade parts classification system *(Parts Type Analysis Team)* **Step 3c** Begin to analyze and revise company policies and practices *(Policy Analysis Team)*	**Step 3**** Undertake a VEP analysis of processes and make reduction proposals *(Process Reduction Team)*	**Step 3** Schedule approved reduction proposals on a VEP Implementation Calendar *(Steering Team)*	
Step 4 Target a priority product as the starting point for Stage 2 reduction analysis *(Steering Team)*			

Stage 2 column spans: *3-View Analysis Teams* (Steps 1a, 1b, 1c)

* *The Control Points Reduction Team works independent from other VEP Teams—and can start its work in Stage 1 (see Chapter Ten).*
** *This team is an optional step, dependent on current levels of operational complexity and improvement.*

Control points always represent negative variety even when the customer requires them, such as with compliance and audit tracking and reports. The result is blizzards of paperwork and computer transactions that pile up in offices and on desks in every corner of the corporation. They are the epitome of C-Costs (Control Costs).

Negative Variety In Processes

As a natural extension of 3-View Analysis, removing a part, product, service or market offering results in the parallel elimination of associated processes. Process reductions are further increased through the work of a *separate and distinct* VEP Process Analysis Team. Such a team is especially important if the company has never addressed the issues of process improvement or cell design. For organizations active in lean or continuous improvement, many unnecessary and redundant processes will have already been eliminated.

Unwarranted variety in processes is not always easy to spot for several reasons. First, processes are typically not well-documented. Few companies have ever taken the time to catalog processes, develop an inventory of them, or monitor the entry of new ones. In addition, process names are not clear. These characteristics are intimately linked.

Using PUI as a case study, let's walk through VEP's procedure for analyzing and reducing processes.

Reducing Processes at PUI

In the first three months of the VEP implementation at PUI, the Early Victories (EV) Team eliminated over 1,000 part numbers. The majority of these were deadwood in the system that PUI meant to remove ages ago—but just hadn't. The EV Team then worked with the shop floor and eliminated over 100 production processes associated with those part numbers—plus 12 dies, 15 fixtures, and 2 machines.

Later, when the Market Analysis Team targeted eight different product series for elimination, more than a dozen associated processes and scores of dies, fixtures, and tools were also removed. (See Chapter Nine, Figure 9.6 Reduction Recommendations.) The other 3-View Teams proposed 65 additional processes for elimination, plus associated dies, tooling, and fixtures. This, in turn, would free up an estimated 1,500 square feet of the production floor and storage space.

Five months into the implementation, the Steering Team formed a separate cross-functional team to focus exclusively on process reduction. The team's target was a 20% reduction from the current baseline. All the right groups were represented: Manufacturing Engineering, Marketing, Operations, Material Handling, Machine Shop, and Stores. As they made their way around the floor, team members saw lots of complexity—negative variety. For example, they saw the same part being inserted four completely different ways. They discovered that certain jobs were sometimes done with fixtures, sometimes not. Assembly jobs were done by hand some of the time and by machine at other times—apparently depending on the quantity ordered and the operator doing the job. No one they talked with seemed uncomfortable with the range of differences—but no one could explain them either.

As team members spoke with operators, supervisors, managers, process engineers, they were also surprised at the many different names they discovered for the various processes and operations. In fact, VEP's first step in reducing processes is to standardize the terminology used to describe them.

First: Standardize the Nomenclature

In some organizations, there are so many different names for a process that operators themselves get confused. It can be even worse for managers. In a finishing operation, for example, the same work may be referred to as: sanding, hand sanding, abrasive processing or grinding; at other times, it may be described generically as "removing material." But "removing material" may—in the same organization—mean a milling operation. In one shop, the same bending operation was variously referred to as "putting right angles on brackets," "angling a part," "bending metal," and "forming corner on metal."

Is it wrong to describe the same things in different terms? Yes! In VEP it is. A torrent of terms referring to the same operation is negative variety in action. You need a classification system—and that requires standardized names. PUI developed one, step by step, in the following way.

- Team members wrote a sticky note for each name on the list they compiled: 142 in total (Figure 10.2).
- They grouped them into category sets and sub-sets.
- Then they brainstormed to develop standard names for each.
- Once they agreed on a single name for each operation, the original list of 142 names had been distilled down to 8 broad process categories and 28 sub-processes or operations (Figure 10.3).

Figure 10.2. **Before VEP: Names for PUI Production Processes**

BEFORE VEP: 142 PUI PRODUCTION PROCESSES

#	Process	#	Process	#	Process
1*	Mounting In Fixture	49	Bonding	96	Scraping Off Rough Edges
2	Dropping Parts Into Fixture	50	Epoxying Parts Together	97	Filing Parts
3	Holding Parts	51	Super Gluing	98	Filing Edges
4	Assembling Parts	52	Adding Loctite	99	Greasing Parts
5	Loading Tool	53	Putting Insulating Material In Enclosures	100	Oiling Equipment
6	Loading Fixture	54	Filling	101	Oiling Component Parts
7	Unloading Tool	55	Potting Switches	102	Adding Lubricant
8	Unloading Fixture	56	Mixing Potting Materials	103	Applying Grease Or Oil
9	Loading Component	57	Epoxy Coating	104	Greasing Threads
10	Unloading Component	58	Spraying Epoxy	105	Putting On Labels
11	Mating Parts	59	Painting On Insulation Material	106	Making Nameplates
12	Inserting Parts	60	Epoxy Option	107	Writing Up Tags
13	Adding Parts	61	Spray Paint	108	Adding Product I.D.
14	Machining Processes	62	Painting Parts	109	Stamping Part Numbers
15	CNC Work	63	Changing Color Of Enclosures	110	Coloring Coding Parts
16	Punching	64	Plating	111	Coloring Mating Parts
17	Punch Press	65	Nickel Plating	112	Marking Machine
18	Arbor Press	66	Soldering	113	Laser Printing For Labels
19	Air Cylinder Punching	67	Arc Welding	114	Computerized Nameplate Machine
20	Drilling	68	Low Temperature Joining	115	Sniping Wires
21	Drilling By Hand	69	Resistance Welding	116	Cutting Back Wires
22	Drilling With Air Drill	70	High Temperature Metal Joining	117	Stripping Wire Ends
23	Drill Press	71	Brazing	118	Wire Cutter Machine
24	Internal Threading	72	Metal Joining	119	Hand Cutting Wire
25	Tapping	73	Spot Welding	120	Wire Cutters
26	Threading	74	Wave Solder Machine	121	Removing Wire Coating
27	External Threading	75	Soldering PC Boards	122	Undressing Wire Cable
28	Removing Material	76	Iron Soldering	123	Splitting Wire Cable
29	Punching Holes	77	Torch Brazing	124	Stripping Fixture
30	Lathe	78	Brown & Sharp	125	Burn-In Test
31	Turning	79	Putting In Rivets	126	Adding Setpoint
32	Milling Machine	80	Riveting	127	Product Calibration
33	Milling	81	Adding Fasteners	128	Testing A Product
34	Turning Down Parts	82	Screwing	129	Calibrating
35	Milling Components	83	Press-Fitting Components	130	Pressure Stand Work
36	Grinding Wheel	84	Pressing Dowel Pins	131	QC'ing Product
37	Hand Sanding	85	Knurled Pin Press Operation	132	Adjusting Setting
38	Removing Material	86	Fitting Machine	133	Test And Calibration
39	Finishing	87	Staking Machine	134	Test Rig Calibration
40	Sander	88	Hand Staking Tool	135	Immersion Bath Procedure
42	Grinding Down Parts	89	Fastening	136	Thermal Chamber Test
43	Abrasive Process	90	Cleaning Parts	137	Vibration Check
44	Angling A Part	91	Degreasing Parts	138	Hose Down Test
45	Bending Metal	92	Wiping Off Oil	139	Up/Down Scale Test
46	Putting Right Angles on Brackets	93	Removing Excess Material	140	Life Testing
47	Forming Corners On Metal	94	Removing Sharp Edges	141	Component Calibration
48	Gluing	95	Tumbling Parts	142	PC Board Functional Testing

* The number preceding each process represents a numerical order, not a code.

Figure 10.3. **After VEP: Names for PUI Production Processes (Grouped by Category)**

PRODUCTION PROCESS	OPERATION	DESCRIPTION
1.0. Assembly	1.1. Loading	Putting in/removing a component from tool or fixture
	1.2. Inserting	Putting a part in on another part(s)
2.0. Machining	2.1. Punching	Removing material using some type of press or fixture
	2.2. Drilling	Removing material via a drill or drill press
	2.3. Tapping	Removing material to create internal threading
	2.4. Turning	Any work done on a lathe
	2.5. Milling	Any work done on a milling machine
	2.6. Grinding/Sanding	Removing material by abrading or grinding
	2.7. Bending/Forming	Re-forming of a part
3.0. Gluing & Finishing	3.1. Adhesive Bonding	Gluing parts together
	3.2. Potting	Filling a cavity with insulating material
	3.3. Coating	Applying insulating coating material
	3.4. Painting	Applying paint to a part(s)
4.0. Soldering/Welding	4.1. Brazing	Metal joining, using filler metal melting above 800°F
	4.2. Soldering	Metal joining, using filler metal melting below 800°F
	4.3. Wave Soldering	Soldering Printed Circuit Boards, using wave soldering machine
	4.4. Resistance Welding	Metal joining, using a spot welder
	4.5. Arc Welding	Metal joining, using an arc welder
5.0. Fastening	5.1. Riveting	Installing and securing rivets
	5.2. Pressing	Assembling press-fit components by hand or by machine
	5.3. Screwing	Installing screws to secure components together
	5.4. Staking	Forming material to secure components together
6.0. Cleaning/Marking	6.1. Cleaning	Removing unwanted material by cleaning, degreasing or wiping
	6.2. De-Burring	Removing unwanted sharp edges by filing, tumbling or scraping
	6.3. Lubricating	Applying grease, oil or other lubricant
	6.4. Marking	Applying any type of identification to a part
7.0. Wire	7.1. Wire Cutting	Cutting wire either by machine or by hand
	7.2. Wire Stripping	Stripping wire coating off by machine or by hand
8.0. Testing	8.1. Testing	Testing, calibrating or setting of a product

Next, the team developed a process *attribute template* for capturing the critical specifications of each remaining process. Similar to the parts attribute template developed in Stage 1 (Chapter Seven), this grid displays the characteristics that need to be considered when describing and cataloging a given process (Figure 10.4). The template also facilitates the entry of accurate, uniform information into the new database.

Figure 10.4. **PUI Process Attribute Template: Welding**

NUMBER	GAS FLOW	WELD CURRENT	TORCH POSITION	TOOL	PART NUMBER	MATERIAL 1	MATERIAL 2
X-31	20-25 PSI	100 Amp	45°	DNV-131	Y-88-A	1/8" SS	1/8" SS
X-37	20-25 PSI	72-75 Amp	90°	DNV-137	Y-253-E	1 1/2" SS	1/4" SS

At PUI, doing this was so painless that a few team members went on to develop a similar listing for tools, fixtures, and dies. With this groundwork in place, the Process Analysis Team was equipped to begin to identify reduction opportunities.

Second: Reduce Processes

Now that the nomenclature is standardized and data entry has been facilitated by using a Process Attribute Template, the Process Analysis Team is ready to deploy the reduction procedure:

1. Target a process category for analysis (for example, soldering and welding).
2. Identify the functions, process specifications, and tooling requirements for this category.
3. Apply the Six VATs and brainstorm possible reduction opportunities, querying:
   - Can this process be eliminated?
   - Can it be combined with another process?
   - Can a new process be created that eliminates this one and others?
   - Are the dies, tooling or fixtures associated with this process unique or can they be shared?
4. Evaluate reduction proposals systematically against stakeholder concerns, holding in reserve any proposal that does not pass this review.
5. Prioritize validated proposals according to the least/most resistance to change and least/most impact on cost.

6. Submit proposals to the Steering Team for review and coordination with proposals from other teams.
7. Choose a new process category for reduction analysis and return to step 1.

As they move through these Stage 2 steps, members of the Process Analysis Team look for factors that cause processes to proliferate, including those related to formal and informal policies and practices. In Stage 3 of the VEP Method, the team will submit their reduction recommendations to the Steering Team for review. Viable proposals are then implemented in Stage 4 and the number of processes decreases. But this analysis team continues to monitor the shop floor for new opportunities and for signs of unwarranted process proliferation.

Between the process reduction associated with the work of the 3-View Teams and the more focused process work of this team, your process universe can shrink considerably. Let's now move on to the analysis of the control point's universe.

Negative Variety In Control Points

A control point is any activity or transaction—paper, electronic or otherwise—that supports the design, procurement, sorting, retrieval, inspection, production, cataloging or handling of parts and products—or their marketing equivalents.

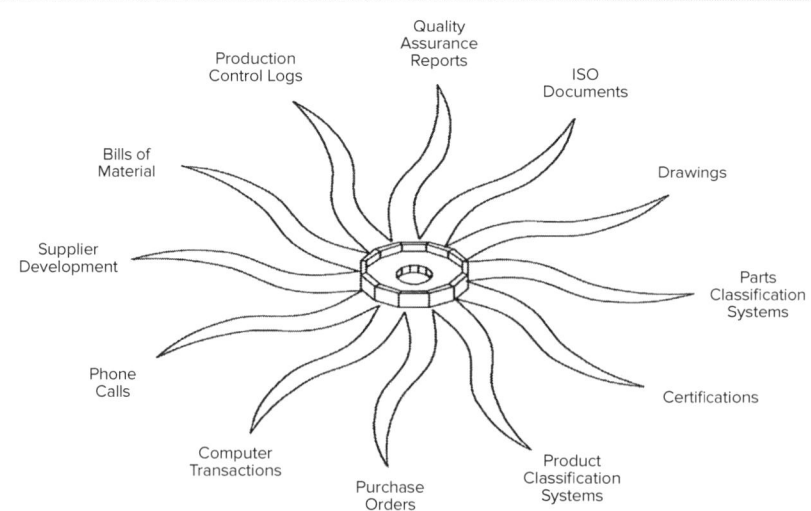

Figure 10.5. **A Part Sprouting Control Points**

Proliferating paperwork and computer transactions are a bane of corporate existence. But companies rarely, if ever, identify or track them in a systematic or comprehensive manner. Yet control points—the vast majority of which are computer-based—consume enormous amounts of an organization's resources of time, money, and human capital. It is complexity that is chronic, rampant, and nearly invisible in accounting terms. The complexity of control points can rob the company of its future simply because few resources remain for innovation or growth.

In manufacturing, it all starts innocently enough—with the introduction of one new part. Let's walk through this progression metaphorically from the control point perspective. Imagine that new part as a seed. Even before that new part is "planted" in a product, it has begun to sprout control points (Figure 10.5).

Although we may prefer to think of each part residing in its own uncluttered little burrow, reality soon makes itself known. The longer the part is "alive," the more control points associated with that part multiply. Joined by other parts and their related control points, the organization turns into a veritable jungle of complexity (Figure 10.6). The congestion spreads slowly and surely, covering the entire enterprise. The more the company struggles to free itself from complexity's grip, the more it becomes entangled.

Figure 10.6. **Control Points: What Really Happens**

We would like to think of each part as residing in its own uncluttered pigeonhole...	Part A	Part B	Part C	Part D	Part E
	Part A	Part B	Part C	Part D	Part E

...but from its outset, each part begins to sprout control points.

As parts and their control points multiply, the jungle of complexity spreads.

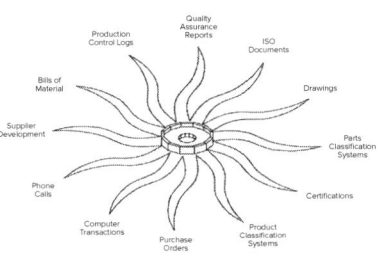
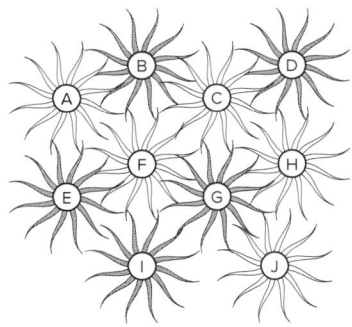

Control points adhere to parts, but they are not always easy to detect. Twenty-six control points, for example, are required for PUI's electronic sub-assembly XJ-889, composed of just three purchased parts and seven made parts (Figure 10.7). This number does not include the myriad computer transactions triggered by that sub-assembly.

If the number of control points reaches 26 for a single sub-assembly, that number climbs exponentially for an entire parts inventory. But let's look more deeply into the nature of the control point challenge. Here's part of the scenario at PUI.

Figure 10.7. **Control Points: Associated with XJ-889 Sub-Assembly**

CONTROL POINTS FOR PURCHASED PARTS: XJ-889	CONTROL POINTS FOR MADE PARTS: XJ-889
Drawings	Drawings
Purchase Orders	Production Orders
Certificate of Compliance	Manufacturing Procedures
Receiving Paperwork Creation	Inspection and Testing
Receiving Transactions	Parts and Fixture Retrieval
Inspection	Computerized Scheduling Transactions
Expediting	Receipt Transactions to Stock
Product Master File Coding	Inventorying
Unpacking	Routings
Paying Suppliers and Freight Bills	Product Master File Coding
Order Status Reports	Moving, Storing
Shortage Reports	Shortage Reports
Packaging to Avoid Damage	Packaging to Avoid Damage

The Search for Control Points at PUI

Often mistaken as an indispensable part of doing business, control points do proliferate when a company is unaware of their negative impact or that they are controllable. Once the company recognizes that control points are at once pervasive and hidden, the hunt is on. To begin your search, let's look at three common places where they are concealed.

Procedures: Documented and Undocumented. At PUI, the number and types of work procedures vary widely from department to department. For example, when the company decided to seek ISO registration—a huge documentation challenge in itself—employees handling paperwork braced for the mountain of effort they knew would be required to document these procedures.

In Operations, as another example, VP Wardwell holds her direct reports responsible for documenting required procedures. But she also realizes that, at this point, no one knows which procedures should or should not be documented. Wardwell recently remarked, "Quite honestly, we do a lot of things without formal written procedures—even though we know the same procedure should be followed each and every time we do something, like building a product, placing a purchase order or 100% testing an outgoing shipment. I think the sheer variety of control activities performed by Operations has a good deal to do with why we do a poor job of creating and maintaining work procedures. Everything seems to be an exception."

Drawings: Standardizing and Maintaining. The number of drawings at PUI and the amount of maintenance associated with them is a significant problem. But no one knows how significant since the true cost of a drawing has never been determined, except perhaps in the payroll figures of the Drafting—CAD/CAM Department.

One cost indicator in the Drafting Department is lead time, which is now measured in weeks, not days, and is a source of frustration for that department as well as for those it supports. There are many days when Drafting Manager, Art Brennan, wonders how many other, more productive things he and his staff could do if the sheer number of drawings was reduced. How often has he discovered staff working on drawings that should have been obsoleted six months ago? How many drawings are there on duplicate parts? According to Brennan, the answer to both questions is: "A lot!"

Forms: Number, Variety, and Overlap. Every PUI department uses dozens of forms in the course of a normal business day and many more at the end of the month. Most forms are designed by individual departments because, managers point out, they have "special" needs. Consequently, offices are awash in purchase orders, logs, fax transmittal sheets, etc., that are all slightly (but, according to each department, importantly) different. In response to a recent customer product query, a customer service person griped: "Should I send the ECN, SRO or MCN form— or all of the above—to get this looked at by Engineering?!?"

The Control Points Reduction Team and Its Tasks

The Control Points Reduction Team is formed earlier in a VEP implementation than the other analysis teams and follows a parallel course. Because control points are linked and also found everywhere in the corporation, this team includes representatives from all departments: Purchasing, Systems, Drafting, Engineering, Materials, Finance, Quality Assurance/Control, Marketing, and Customer Service. Its mandate is to:

- Classify and analyze control points by department and company-wide, applying the Six VATs to identify viable reduction opportunities.
- Recommend revisions in policies and practices to better regulate and minimize the addition of control points in the future.

As you may already anticipate, the team begins by developing a classification system for control points—if one does not currently exist. This, of course, requires that:

- The terminology related to control points (names of forms, transactions, etc.) is simplified and standardized.
- Control points are identified and cataloged.

To do this, the team selects a reduction focus—in all likelihood forms and reports (order entry, credit applications, accounts receivable activity report, accounts payable activity report, deposit slips, etc.). Next, with the help of their departmental colleagues, individual team members compile hard copies of all forms used in their respective departments and determine how often each form is used (daily, weekly, monthly, quarterly, annually). The forms are then grouped into categories, according to underlying commonalities, and a label is found for each group—reports, fax forms, order forms, accounting documents, and so on.

The next step is to put each group of forms into a master binder by category and to develop a master list for the department. This master list is put into a matrix, showing each form and its usage frequency, so that an index can be calculated. This index—a quantification of the relationship of form type and use—is exactly parallel in form and function to the VEP Parts Index used in the 3-View Analysis process (Figure 10.8). The result—the number of forms times the frequency of their use—is the indexed figure.

The team is now ready to reconvene, with each department bringing its binder and index to the meeting so that a company-wide binder and a comprehensive index can be further developed.

Using the company-wide index and hard-copy master as baselines, the team now begins the analysis process. The aim is to find ways of simplifying, combining or eliminating forms, applying the Six VATs. The

Figure 10.8. **VEP Control Points Index: Accounting Forms**

	NAME OF FORM	A DAY (X 252)	B WEEK (X 52)	C MONTH (X 12)	D QUARTER (X 4)	E 6 MONTH (X 2)	F YEAR (X 1)	OCCUR- RENCES
1	5-Days-In-Receiving Report		1					52
2	A/P Aging Report			1				12
3	A20	6*						1,512
4	Account Summary By Cost Center		1					52
5	A/R Activity Report	1						252
6	AR/Aged Trial Bal Detail	1						252
7	Application for Credit		1					52
8	Authorization for Payment		4					208
9	Back Order Stock Ship List	3						756
10	Bank Change/Info on Address	1						52
11	Bank Deposit Receipt	1						252
12	Bank Deposit Ticket	1						252
13	Buy-Part P.O.		1					52
14	Canadian Pick List	3						252
15	Cash Deposit Report			1				12
16	Cash Receipts	1						252
17	Cash Receipts Final Register	1						252
18	Certification Approval Notice				4			16
	Total Discrete Forms	10	5	2	1	0	0 18	5,044

VEP Index for Accounting Forms = 90,792 (18 forms x 5,044 total occurrences)

Numbers in columns A through F indicate the number of distribution points (people) for each form within the various time intervals.

first three tools (Unique vs. Shared, Modularity, and Multi-functionality & Synthesis) will prove most helpful, with the last three (Ease of Assembly, Range, and Trend) having more limited applicability. (See Chapter Eight for a review.) Some specific questions related to forms are:

- Can this form be eliminated?
- Can this form be combined with another existing form?
- Can a new form be created that eliminates this form and even other forms?
- Can this form be revised so that it eliminates the need for another form?
- Can the information on this form be communicated or retained in another way—for example, computerized for tablet or PC?

Figure 10.9 is an example of an improvement work sheet one company developed for this purpose.

Figure 10.9. **Sample: VEP Improvement Work Sheet for Forms**

VEP Improvement Work Sheet for Forms (Control Points)

Your Name _____   Today's Date _____

Department _____   Dep't Head _____

Name of Form _____

Use Frequency (circle as applies)	Daily	Weekly	Monthly	Quarterly	Twice yearly	Yearly
	252	52	12	4	2	1

1. Please check as many of the following boxes as apply:

   This is an **internal form** (used in your department) ☐
   This is an **external form** (used in other departments) ☐
   This is an **outside form** (used in outside the company) ☐

2. ___ **No Change is proposed for this Form.** Please explain why: _____
   _____

3. ___ **Changes are proposed to this Form.** Please check and explain as applies (Staple related forms to this one.)

   a. ___ This Form can be eliminated _____

   b. ___ This Form can be combined with an existing form _____

      • Name the Existing Form _____

   c. ___ A new Form can be created that eliminates this form _____

      • Name any other form that can be eliminated _____

      • Name the new form: _____

      • Please attach a draft of the new form _____

   d. ___ This Form can be revised to eliminate another form _____

      • Name of other form that can be eliminated _____

      • Name the new form: _____

      • Please attach a draft of revised form _____

   e. ___ Other Proposed Changes _____

___Approved   ___ Approval Pending   ___Not Approved (Here's why: _____
_____

Implementation Completed Date _____
Remarks _____

Interface with the VEP Steering Team. While the Control Points Reduction Team keeps the Steering Team closely informed of its progress, it is more autonomous than other VEP teams since there is little interface with parts per se. Decisions, therefore, can be made by consensus, linking directly back to other departments as necessary. The team is also empowered to implement its improvement and reduction proposals without Steering Team approval.

Predictably and similar to the other teams, the control points group is on the constant lookout for internal and external causes of control point proliferation and ways to eliminate these causes. Once all remaining control points are validated and put on a master list, team members regularly monitor that list for any signs of proliferation. Gatekeeping functions can be set up in each department to monitor this activity.

Actual Results. Members of the Control Points Reduction Team in one company were very proud of their results. Having identified 226 forms, they eliminated 52 within the first 90 days of activity. Six months later only four new forms had been added.

More impressively, they reduced the Forms Index by nearly 20,000. Control points reduction in the Drafting Department was also dramatic. Seventy-two cabinet drawers of drawings were whittled down to sixteen—an 80% reduction (Figure 10.10).

Figure 10.10. **Actual Control Point Reduction Results (Partial)**

CONTROL POINT	BEFORE	AFTER	ADDED SINCE
Actual Number of Forms	226	174	4
Forms Index	62,449	43,025	548
Actual Drawings	72 cabinet drawers	16 cabinet drawers	1 cabinet drawer

Downstream Is Where The Silt Piles Up

That single new part is an almost-invisible obstruction in the flow as products make their way to the end-user. But look downstream from that part for further evidence of its impact. There you see a silt pile-up in the form of excessive processes and excessive control points. These function like a dam in the flow, slowing it down to a trickle.

Look downstream, too, for a gauge of your progress in regulating mushrooming parts. As unwarranted parts, products, and processes are removed, the flow gains speed, getting progressively more rapid with the elimination of unneeded control points. The success of your efforts to curb and minimize processes and control points is reflected in a new sense of spaciousness and ease in the company's internal systems. The infrastructure regains some breathing room and, as a result, the flow accelerates.

CHAPTER 11

Implementing VEP Improvements

> VEP is strategic in nature, requiring high-level support with a vision for simplicity and quality. When truly practiced, it's a game changer for your company and your industry.
>
> — ERIC LIAL, *Transportation Insight*

The heart of the VEP Methodology is the 3-View Analysis Process, with its penetrating focus on Parts Type, Product Structure, and Market Offerings—and secondarily on processes and control points. Critical though this analysis is, of equal importance is what comes next in Stages 3 and 4:

- Reviewing and prioritizing the reduction recommendations these teams submit to the Steering Team (Stage 3).
- Implementing these priorities and taking concrete steps to prevent the reoccurrence of negative variety (Stage 4) (Figure 11.1)

In this chapter, after a quick review of Stages 1 and 2, we un-nest Stages 3 and 4, explaining how each helps the company harvest the hard work already in place. At the close of the chapter, we discuss high-level implementation issues for your consideration.

Stage 1 Review: Prepare

Stage 1 begins with determining the scope of your VEP initiative: deep-dive or discrete. Either way, this first VEP stage consists of vital preparation: setting up and training a task force or a full complement of VEP teams and choosing your target product line. If you choose a deep-dive approach, Stage 1 includes making your launch official, rolling out VEP Awareness Sessions for all employees, and getting some early victories under your belt. Either deep-dive or discrete, Stage 1 also requires that

Figure 10.1. **VEP Methodology: Stage 2/ Reduction Analysis for Processes and Control Points**

STAGE 1 Plan and Prepare for an Effective VEP Implementation	STAGE 2 Identify Reduction Opportunities by Applying the Six VATs	STAGE 3 Coordinate and Schedule Reduction Proposals	STAGE 4 Implement Improvements and Sustain a VEP Mindset
Step 1 Select a Steering Team that then sets up the other VEP Teams *(Management)*	*3-View Analysis Teams* **Step 1a** Undertake a VEP analysis of market offerings and their characteristics and make reduction proposals *(Market Analysis Team)* **Step 1b** Undertake a VEP analysis of parts as part of the product architecture and make reduction proposals *(Product Structure Analysis Team)* **Step 1c** Undertake a VEP analysis of parts by parts type and make reduction proposals *(Parts Type Analysis Team)*	**Step 1** Coordinate and consolidate reduction proposals *(Steering Team)*	**Step 1** Implement approved reduction proposals *(all VEP Teams)*
Step 2 Conduct VEP training for teams and begin general awareness training *(Education and Methods Team)*	**Step 2*** Undertake a VEP analysis of transactions that support parts, products, and market offerings and make reduction proposals *(Control Points Reduction Team)*	**Step 2** Qualify, approve, and prioritize reduction proposals *(Steering Team)*	**Step 2** Set up a VEP Preventative Monitoring Calendar and continue to educate a VEP mindset *(all VEP Teams)*
Step 3a Find and reduce parts with low-resistance to change *(Early Victories Team)* **Step 3b** Assess, clean up, and upgrade parts classification system *(Parts Type Analysis Team)* **Step 3c** Begin to analyze and revise company policies and practices *(Policy Analysis Team)*	**Step 3**** Undertake a VEP analysis of processes and make reduction proposals *(Process Reduction Team)*	**Step 3** Schedule approved reduction proposals on a VEP Implementation Calendar *(Steering Team)*	
Step 4 Target a priority product as the starting point for Stage 2 reduction analysis *(Steering Team)*			

* *The Control Points Reduction Team works independent from other VEP Teams—and can start its work in Stage 1 (see Chapter Ten).*
** *This team is an optional step, dependent on current levels of operational complexity and improvement.*

you undertake the task of assessing, upgrading, and revamping your parts classification system. This task is arguably the most complex of Stage 1—and it is crucial. Yes, this activity yields a VEP-capable classification system that serves as the foundation for the analysis to come—and subsequent improvements.

The discussion that follows all assumes the full deep-dive scope.

Stage 2 Review: Analyze By Applying The Six VATs

In Stage 2, the newly-formed VEP teams begin their work. The Policy Team begins its active search for the formal and informal policies that trigger unwarranted variation. The three 3-View Analysis Teams (Market Analysis, Product Structure Analysis, and Parts Type Analysis) apply the six VEP analysis tools (Six VATs) to find, investigate, and minimize negative variety. Synergies between these teams strengthen their analysis further. Often in close parallel, the Control Points Reduction Team utilizes the Six VATs to root out the transactions that adhere to each part, product, and market offering. Sometimes a fourth team undertakes Process Reduction, though most companies shrink processes in response to 3-View activity. All teams surface a stream of improvement proposals and send them—on a steady and on-going basis—to the Steering Team for review, approval, and implementation. As Stage 2 continues, policies and practices change in keeping with VEP principles. VEP thinking starts to become a part of the company's improvement mindset.

Stages Three And Four

The final two VEP stages are closely aligned in purpose and intent: to review, approve, and prioritize VEP improvement proposals; next implement them—and then take concrete steps to prevent the recurrence of negative variety.

Stage 3: Prioritize and Schedule

Picture, if you will, improvement proposals flowing into the Steering Team from three sources—the Market Analysis, Product Structure Analysis, and Parts Type Teams. Proposals are also submitted by the Process Reduction Team if such a team has been separately designated. Remember, the

Control Points Reduction Team typically works independently, in parallel with the 3-View Analysis Teams.

As these improvement proposals flow into the Steering Team, team members review and coordinate these reduction ideas, which range from opportunities in parts, products, and overall market offerings to needed policy revisions and additions. Always alert to proposal consolidation, the team also keeps a keen eye out for similarities, overlaps, and redundancies as well as for outliers and anomalies—which can be considerable.

For these reasons and more, the team must develop a sturdy review process. Figure 11.2 shows the *VEP Proposal Impact Worksheet* that one Steering Team used to prioritize each improvement proposal by assessing its impact on a set of factors this company considered primary—from agency approvals to customer acceptance. The same proposal is then evaluated against a set of secondary factors, ranging from volume to expected qualitative benefit. Also notice the column for assessing the proposal against F-, V-, and C-Costs—true cost. This is a worksheet well worth your study.

While this can sometimes be aided by software, PUI's Steering Team opted for a low-cost sticky notes approach. The process was simple and highly effective: each proposal was written on its own note, posted on a length of brown paper, and moved around until a proper relationship with the other proposals was found. Because this approach was so concrete and hands-on, the company reserved an entire room to serve as the Steering Team base camp.

Yes, reviewing and processing improvement proposals is work—but this is work that is spread across an extended period of time. Remember, the Steering Team has been meeting since the VEP launch, vetting recommendations as they come in. In addition, submitted proposals are already pre-qualified to some extent since each must be accepted by relevant stakeholders *before* it reaches the Steering Team. This reduces the number of proposals needing careful review.

Taken in sum, the Steering Team has flexibility in approving and prioritizing proposals. Theirs is not a lock-step procedure. For example, the Steering Team can and often does encourage teams to implement certain proposals long before Stages 3 or 4—sometimes even in Stage 1. Proposals characterized as having low resistance to change, for example, or requiring minimal resources to implement can quickly move through to implementation. In keeping with the goals of the Early Victories

Figure 11.2. Sample: VEP Proposal Impact Worksheet

VEP Improvement Proposal: Impact Work Sheet

Product Type _____ Date _____

Part Number _____ ___Fixed ___Quasi-Variable ___Variable

Core Function _____

Attribute _____

Core Question _____

Mark the Impact that Each Proposed Change will have on Each of the Following, Using a Scale of 1-10 (10=high).

VEP Improvement Proposal	Prioritize	Mark as applies: F-Cost V-Cost C-Cost	1 Fixturing	2 Assembly	3 Product Function	4 Agency Approval	5 Liability	6 Field Service	7 Customer Acceptance	8 Volume	9 Costs	10 Operations Change (+/-)	11 Expected Benefit (Qualitative)	12 Foreign Distributors
1														
2						PRIMARY					SECONDARY			
3														
4														
5														
6														
The four-dot system is used to process the proposed ideas ➡			● OF INTEREST	●● UNDER PREP	-●●- NOT VIABLE	●●● READY TO TEST	●●●● TEST	●●●● DID NOT TEST WELL: NEEDS MORE WORK OR REJECT	●●●● SUCCESSFUL TEST: READY TO IMPLEMENT	●●●● IMPLEMENTED				

$$$ The Cost Of Making The Change

Team, these early reductions garner special visibility and do a double service—first, for their improvement impact and, second, for their positive promotional effect.

Nevertheless, some ideas need to wait for Stage 3. They may

represent a costly, complex change that cannot be undertaken lightly. Other improvements may require a high degree of coordination and/or multiple approvals. In Stage 3, these proposals are prioritized and entered into a formal implementation calendar.

There is wisdom in this conservative pacing, especially if teams are still working through the product and parts universe. One excellent multi-functional idea may surface, for example, in the early days of product structure analysis, only to be superseded a month later when the Parts Type Team finds an even wider application. The reasoning is this: First go with the changes that are low-cost and can ease through the system—and exercise patience on the rest.

This waiting is not time lost. You'd be amazed (or maybe you wouldn't) at the depth of improvement that can transpire in the values and day-to-day procedures of a workforce actively engaged in VEP. Awareness of VEP principles and goals and the intention to realize them can inspire people to make simple improvements on their own, the kind that never reach any VEP team. This is especially so when coupled with concrete examples close to home. Because people are changing the way they think, they seek ways to make those changes concrete.

We have seen this in industrial design and product engineering groups who already want to do it right: save time and cost—and make the company money. Once their consciousness is raised about effective variety, people find their own ways to get on with it. Practices around product development, for example, often align quietly—but decidedly—with a new level of VEP awareness and intention. As a result, incremental gains begin to build. These informal gains fuel the implementation and keep the momentum and excitement alive as teams comply with the more formal, step-by-step procedure of the VEP Methodology. These informal gains are also an important part of the method.

Stage 4: Implement and Prevent

At the start of Stage 4, many key VEP proposals have already been implemented or are prioritized on the Steering Team Calendar. This prioritization allows changes to enter the company at an even and sensible pace, in doses that do not overload the system. In this regard, Stage 4 resembles a massive ECN (engineering change notice) process. Given that all VEP proposals are pre-qualified and represent exceptionally clear thinking, this process can strengthen—or even replace—a traditional ECN approach. In any event, the balance of the VEP improvement proposals is deployed in this stage.

This fourth stage of the methodology, however, has another dimension: prevention, one of VEP's three main goals. By way of reminder, those goals are: 1) maximize customer selection; 2) reduce negative variety; and 3) prevent the recurrence of negative variety. A VEP implementation cannot be considered a success *until and unless* the roots of negative variety are dug out of the organization and replaced by practices and policies that prevent it from returning.

Once again, all the VEP teams are involved in this important work. Their ideas can be imaginative, as seen in the notion of "border guards" to watchdog against unwarranted additions to the parts base. They can be practical—as in the company that decides to weed out all the deadwood in their drawing files and remove all the now-emptied cabinets so new drawings could not find easy places to hide. These ideas and more are identified and implemented in this stage.

Finally, the Steering Team creates a special VEP Review Calendar that schedules regular scrutiny of the negative variety indicators that members of other teams have discovered, companywide. Each team watchdogs its own point of focus. For example, insiders to the engineering function become vigilant about additions to the parts type inventory, using the set of attribute templates developed in Stage 2 as a routine part of their qualifying procedure. Similarly, using the Six VATs to search for unwarranted variety in product architecture becomes a habit of mind for all analytical teams. Members of the Control Points Reduction Team, for example, continue to meet quarterly to review new forms in their respective departments. Everyone becomes vigilant policy-monitors. Negative variety doesn't stand a chance in this informed work culture. VEP improvements stabilize and grow.

Your first VEP implementation cycle is likely to take from 12 to 18 months. If your organization is particularly challenged, a full cyce of VEP improvement could well take an additional 6 to 9 months. We include a sample 12-month schedule for you to use as a starting point for preparing your own timeline (Figure 11.3). With that, we round out our discussion of the four stages of the VEP Methodology and their associated principles, tools, and procedures. A rigorous application turns the eradication of negative variety into an action plan that results in double-digit savings and a fresh future for enterprises too long burdened by complexity they could barely identify and never reduce.

Making The Change: Some High-Level Implementation Issues

If yours is a naturally hardworking company, driven by the puritan ethic and a passion for challenges, you may be raring to take VEP on comprehensively—to "go for it." If so, VEP will work you hard—but the methodology will keep you focused and the momentum of improvement will build, one proposal after another. Plugged into VEP's systematic approach, you and your company will experience early satisfaction from a vigorous attack—and, later on, huge cost savings and dramatic jumps in profit margins.

The willingness to "go for VEP" is as much a decision as a state of readiness. The best way for VEP to happen is for company executives to commit to a systematic approach, budget the resources, and publicly state, "We're gonna do this!" Then they need to stay visible. High-profile commitment gets and keeps others on board. If your company's thought-leaders and visionaries have accepted the need for the kind of change VEP is designed to create, you are on your way. And if you happen to be a leader in your company, remember to turn over *authority* to the teams when you turn over *responsibility*.

If, on the other hand, managers hesitate in the face of "all that work" or don't "see" parts variation as a problem—but you do—start with a small pilot on a closely defined product line. Let VEP work its magic within that constraint. Adopting a *slow-and-grow* (discrete) approach can help you build your credibility and support. With those in play, you are in a better position to extend your effort into the next and wider cycle.

Whether you go slow or fast, a key to your success is recognizing that people and organizations can absorb only so much improvement, so much change. This is a crucial understanding. Determine your own personal absorbency level and that of your group or company before you commit to either of the above VEP approaches. In the here-and-now, ask what level of improvement your group can actually handle. Leave the home-run scenarios at the door. Stay rooted in your *actual* reality—your reality now.

Even so, your organizations may not be able to connect with the VEP challenge at this time. In some cases, this will be due to a lack of corporate will. While the company may, on some level, know that the only way to rid the system of complexity, congestion, and unwarranted cost is through a deep VEP clean up, management may simply *not want* to make that level of effort.

Figure 11.3. **Sample: 15-Month VEP Implementation Timeline**

SCHEDULE	VEP TEAM ACTIVITY	
Month 1	1.1.	Senior Management Briefing: Course to orient Senior Managers (1.0 day/30 participants).
	1.2.	Planning: Draft implementation timeline; select Steering and other VEP Teams; and schedule training events (1.0 day)
	1.3.	Technical Training for VEP Teams: Course to instruct all teams in VEP Methodology (3.0 days/30 participants).
Month 2	2.1.	Steering Team qualifies targeted series candidates.
	2.2.	Parts Type Analysis Team begins work on classification system overhaul.
	2.3.	Early Victories Team begins.
	2.4.	Market Analysis Team begins.
	2.5.	Product Structure Analysis Team begins.
	2.6.	Education & Methods Team sets up a VEP Resource Center and schedule for on-going VEP Awareness Sessions for general workforce.
Month 3	3.1.	Awareness Sessions begin for other employees (3.5 hours/30 participants per session).
	3.2.	Steering Team begins active coordination.
	3.3.	Parts Types Analysis Team continues work on making classification system VEP-capable.
	3.4.	Early Victories continues.
Month 4	4.1.	Control Point Team begins.
	4.2.	VEP Policy Team begins.
	4.3.	Early Victories Team concludes its work.
	4.4.	All other VEP Teams continue.
Month 5	5.1.	All VEP Teams continue.
Month 6	6.1.	Parts Type Team completes enough of its overhaul of the parts classification system to begin its parts type analysis.
	6.2.	All other VEP Teams continue.
Month 11	11.1.	Parts Type Team completes its overhaul of the parts classification system and completes its parts type analysis.
	11.2.	All other VEP Teams continue.
Month 13	13.1.	Steering Team finalizes and prioritizes improvement proposals.
Month 14	14.1.	Steering Team undertakes implementation of approved recommendations—all other VEP Teams assist.
Month 15	15.1.	Steering Team completes implementation of currently approved recommendations, with all other VEP Teams assisting.
	15.2.	Steering Team leads the sustainment of a VEP Mindset, setting up "border guards" and mentors in all functions (all other teams assist).

There can be valid reasons for this. After years of effort and commitment to quality and productivity improvement, the company may be suffering from improvement fatigue. Lean, visuality, TQM, kaizen, *poka-yoke*, SMED, Shingo, Six Sigma, teams, ISO certification—the list of improvement interventions grows longer every year. While the vast majority of such methods are good and many are excellent, the notion of "yet another improvement program" can be overwhelming. Assess your company dispassionately in this regard and do not ignore what you find.

The reality is that new improvement approaches are always waiting to get implemented in companies that may already be over-implemented. Companies that have become lean organizations are proud of it. But that's the point: Can a lean organization absorb yet another system, no matter how "right" it is? Can a maxed-out staff deploy yet more improvement? What does a company do when it is so lean that it can't absorb the very solution it knows it desperately needs? Such a company must wait.

Other companies drawn to VEP are in their second or third generation of improvement and looking for the next stretch goal. They have done quality. They have done teams. They have done lean. For their next step, they want a process that is sufficiently congruent with existing programs to make an extension into it natural and logical. They can afford to be aggressive—they have the infrastructure and they have the margin. They have no intention of dropping their current efforts but look, instead, to weave the new into the existing—building on the strengths of both and keeping the improvement momentum going.

No matter what, it is rarely advisable to introduce exciting new improvement methods like VEP with fanfare. Big programs making big promises can result in big disappointments. After the initial buy-in and groundwork are in place, do a low-key launch. Begin by building a common awareness. Introduce employees to VEP principles, concepts, and models of variety effectiveness. Make sure to conduct awareness sessions for as many functions as possible. You may discover, for example, that you get a big boost from involving hourly personnel who have to deal with the day-to-day effects of unwarranted parts variation in the form of congestion, frustration, searching, mistakes, changes, and delays.

Someone once said, "It is a mistake to ignore culture—but it is a worse mistake to make it the principal target of change." Culture is not the enemy—non-aligned behavior is. A company cannot order its culture to change. But it can introduce new ideas that change people's understanding and their mindset. New behaviors emerge out of that, and in response—the culture shifts. In presenting the VEP principles and practices, this book provides the enterprise with a template for that change in understanding.

In the next and final section of this book, we re-cap the major issues imbedded in adopting and adhering to the VEP mindset and making products for profit. There are many benefits to VEP—and they all focus on letting your company secure a genuine and long-lasting competitive advantage. And if you are convinced that VEP can be instrumental—even revolutionary—for your organization, but you are not one of the executives who can make VEP a corporate mandate, the next chapter (Chapter Twelve) is the one to share. Show it to them. They will get it.

PART III

The Bottom Line

CHAPTER 12

CHAPTER 12

Designing For The Bottom Line

Many companies pay for their products twice—once in the development phase and again in the level of complexity they add to their systems.

Organizations are complex organisms in which a great many independent elements interact in a great many ways. During the course of a single workday, literally thousands of individual acts and discrete transactions transpire in support of the company's purpose. These transactions can either weave a rich and successful tapestry of collective effort—or a strangling web of complicated minutia and organizational congestion. Because they are dynamic, organizations can dance along the edge that separates the viable from the complicated—for a while.

Organizations are also flexible and adaptive. They learn. They learn, for example, to cope with the stress that enters with the introduction of each new part, nudging their internal complexity quotients, however fractionally, just that much higher. The fact is: Corporate structures can absorb huge amounts of complicating stress. But just as stress in the body can lead to actual physical complaints—headaches, ulcers, heart disease—sooner or later the burden on a company's infrastructure exhibits itself as assorted and costly problems: long lead times, defects, rework, scrap, a tangle of processes, miscommunication, schedule changes, delays, searching, waiting, soaring costs.

At a given point, the system reaches overload and goes tilt. Stress factors and their compensating responses have become entrenched and the company enters a state of organizational complexity that becomes chronic. Profits erode.

Organizational gridlock and profit erosion are high prices to pay for product diversification. From VEP's perspective, however, the culprit is not the diversification of products. The culprit is *unwarranted* variation and the complication that can follows: the cost of complexity. Even when your hot-selling products reach the end of their productive life and are retired, even when parts are obsoleted and removed, the complexity that surrounded them remains—a tight web of effect and cause that has taken on a life of its own. This web does not come unraveled by itself. It must be *systematically* dismantled, strand by strand.

VEP offers an orderly strategy for this. When the mandate for VEP is set as a strategic priority, coming from the top, the organizational and bottom line impact can be enormous. While attacking waste on the shop floor—or in the front office—affects 5%-10% of the business, a change in top-level improvement thinking affects virtually 100% of the company. That's where the VEP mindset needs to start and continuously reside in order to fully benefit from the methodology set out in this book.

Smart simple design means getting smart about getting simple. The payoffs can be remarkable. Imagine that the principles and methods of variety effectiveness discussed in this book are in practice in your company. Imagine your operating profit increased by 3 to 5 percentage points on average—as recently reported in an A. T. Kearney analysis. Imagine your SKU portfolio reduced by 20% to 40%, your parts levels at 30% to 40% less than those of your competitors—even while you turn out one great new hot-selling product after another. Imagine a development process that is so robust and fluid that product lead time becomes a genuine commercial advantage. And imagine an internal system so efficient in both process and support that your employees, increasingly, have the time and the spaciousness to add real value to their work—and also to innovate. This is the vision of VEP: Variety Effectiveness Process.

This vision is what many companies are in the process of realizing. When Nissan's profits began a downward slide in the 1990s, the company knew it was in trouble. Its response was to look inward instead of outward. The nub of the issue was ineffective variety, coupled with an untenable cost structure. Nissan launched an internal revolution whose goal was: Get simpler. And it did. Instead of needing over 6,000 fasteners across 100 different vehicles, Nissan cut that number to 1,000. In one carline, Nissan slashed floor carpet options from seven to two, steering wheel choices down from 87 to 10, and model selection from 23 to 10.

The direct cost savings of these and other reductions triggered during this period of reform were significant. And the indirect cost savings throughout the company in terms of support functions and organization complexity were inestimable. Nissan became a force to be reckoned with. Witness Nissan's US introduction of its made-in-Japan Maxima. With a base sticker price of $2,500 less than the previous version, this was an astounding corporate feat—not just in terms of comebacks but in the face of an appreciating Japanese yen.

Hewlett-Packard is another company that profited by de-complicating its products and its organization. HP Loveland (Colorado) designs, manufactures, and markets a variety of electronic products for use in the test and measurement market. Its digital multimeter product line, which ranges in price from several hundred to several thousand dollars, represents a multi-million segment of a $5 billion market. But with the de-escalation of military spending, that product series was characterized by slow growth and increased global competition. The Loveland division, however, set its sights on delivering a multimeter (model 34401A) with a performance comparable to that of instruments costing $3,000 to $5,000—but at a price tag of $1,000. Less than two years later, it did!

Using assorted but *linked* development tools (QFD, DFM, DFA, activity-based costing, concurrent engineering), the team designed a product with:

- Reduced assembly time—from 20 to 6 minutes
- Reduced number of hand-assembled parts
- Reduced overall parts count
- Reduced number of screw parts types—from 25 to 2
- Reduced number of required screws—from 27 to 11
- All screws the same size
- An innovative terminal block that snaps into place
- Auto-calibration instead of manual-calibration
- Nearly $200 in reduced direct cost

The product's simplified and reliable design—a finalist for an APEX Product Development Excellence Award—also had a sizeable impact on downstream processes, as evidenced by sharply reduced floor space requirements and inventory levels that dropped from the 60 to 75 day range typical for such instruments to 2 to 4 days. Plus 15 months after production release, not a single ECN on this product was written. Fifteen years later, the 34401A still sells for just over $1,000.

United Electric Controls Company (UE), the company where we piloted variety effectiveness and the VEP Methodology, had similar success. First, having learned the importance of true cost, UE adopted a policy to market *existing* products first, *before* taking steps to broaden its line. As a result, products once thought "too costly" achieved a resurgence in volume—which, in turn, produced increased economies of scale and a better utilization of available production resources.

Secondly, new product development began to focus more clearly on *new* products rather than revised versions of older products. Prior to VEP, engineering resources were frequently tied up developing product extensions which ultimately cannibalized existing sales—and, at the same time, increased variety of parts. The products UE develops today typically fill a new niche rather than overlapping current sales. As a result, engineering resources are deployed with a greater assurance of return.

Finally, attempts to manage costs on new and existing products have proved more fruitful since the advent of VEP. Consideration of true costs, once again, led to more robust designs with fewer parts and processes. The impact of this on variety costs has produced important economies of scale, leading to more inventory turns and improved gross profit at UE. Products developed since the introduction of VEP have 20% to 60% fewer parts than older products. Prior to VEP, conventional wisdom equated variety with customer selection. UE realized that proper management of variety provides the customer with a greater selection of end products and services—while the universe of component parts and processes is continually reduced.

But do not restrict the application of VEP principles and practices to marketing, products, and parts. The need to simplify—to systematically de-complicate—is pervasive and its benefits impressive, no matter the venue or business scenario. Nypro, leading global provider of precision plastic products, provides a radical example. At one point in its history, the company had over 800 customers and was sinking fast. Two years later, Nypro realized that 80% of its profit was coming from 20% of its customers—and "fired" the unprofitable 80% segment. Only forty customers remained, allowing the company to consolidate and re-direct its resources. The strategy paid off. In addition to remarkable cost savings, the other side of Nypro's financials told an even better story. For 14 years, year-on-year, Nypro posted record sales and record profits.

Nissan, Hewlett-Packard, United Electric, and Nypro—all learned to get smart about getting simple and to take it to the bank.

Variety Effectiveness—A Unified Approach

VEP is not a hard sell for companies where proliferating products and parts inventories have become a burning issue. Resources are tight and launch activity is high. To them, variety effectiveness makes sense. No company wants to "reinvent the wheel"—the latch, knob, cover, screw, switch assembly, face plate or engine. They know that any time they avoid re-invention, unutilized resources can be harnessed for other gains, and the results go straight to the bottom line.

In fact, many companies are engaged in some facet of the VEP process—without even knowing it. Some upgrade their classification systems and parts types analysis through Group Technology. Others simplify their product structures and facilitate assembly operations through value analysis/value engineering and through Design for Manufacturability and Design for Assembly. Still others cut their parts inventory and reduce their processes through cell design, one-piece flow, and other lean methods. Many of these businesses are also minimizing control points through paperwork simplification.

Yes, many companies are doing many things to improve their products. But few tie all of them into a single unifying strategy: VEP. Because such efforts are not united, their full value and benefit elude the enterprise. Furthermore, some companies worsen the situation by applying excellent improvement techniques in isolation. An already-mentioned case in point is the fragmented use of value analysis/value engineering for reducing functional costs on individual products while unintentionally triggering variety costs across the product universe.

Any improvement effort done in isolation cannot maximize its potential value to the organization. The price of a non-integrated/non-unified approach is exorbitant and needless cost. Nick Vanderstoop, synchronous administrator for the Canadian division of one of the world's largest automakers, discovered a 40% redundancy across 240 development projects; that is, multiple projects had the same developmental focus. Though the duplication was usually corrected once discovered, Vanderstoop was still moved to observe: "Products, parts, projects! They sprout in this place like fins on a leviathan. Some of them propel you forward. But lots of them paddle in the opposite direction—backwards. When you get too many going backwards, the company is heading for a dead-end and, most of the time, doesn't know why. It's just too darn big to keep track of everything that's happening on its huge body."

Without a unified approach, efforts may lead to improvement but the results are isolated and often absorbed by the remaining noise in the system. No improvement synergies are achieved. Positive impact is minimal.

A similar need for unity recently surfaced in the engineering department of a wildly successful electronics company on the West Coast. For over a decade, this company, with over sixty thousand employees and billions of dollars in annual sales, cashed in on huge cost savings by simplifying its product structures through DFMA technologies. Its leading-edge products continue to win customers and awards. But an Engineering VP confided that, despite an outstanding product track record, the company lacked an overall and unifying approach to product expansion. There was no formal, behavior-based system to ensure that the company's sizeable DFMA investment would be maximized.

The same VP commented on the matter of carrying over/sharing parts. "Whether parts get carried over still depends on which engineer is in the lead or what the preferences of the team are. It's never a matter of routine—or policy. We just don't have any formal practices in place to get us to do it. And we also haven't learned how to leverage off someone else's work in a systematic, disciplined way. Not yet, anyway.

"We are an engineering-driven company. Creativity is in our blood. But we are also highly de-centralized. Everyone likes to do their own thing. That's what they are used to. Yes, we're heavy into DFMA and it's working great. But there really isn't anything to prod us to look at the issue of variety effectiveness from a wider perspective than just the product sitting in front of us. And too few of us see that as a problem."

What other opportunities may this flourishing company be missing—or not maximizing? What is the genuine profit potential of the organization?

Indeed, many organizations have not yet learned to leverage the positive efforts they are making. The policy and marketing arenas of the enterprise are particularly neglected. It is equally rare to see specific improvement practices linked up and integrated, allowing the organization to move from strength to strength. Without a unified and systematic approach, companies can miss what they have nearly in their grasp—the payoff.

But there is reason for hope. More and more companies are beginning to view themselves holistically—as a unity of effort—and not merely as a container for the disparate functions called departments. They know that it is not enough to be linked together by demand. Because organizations are dynamic systems, improvement efforts must be connected to the larger issues of the organization as an aggregate.

The Goal Is To Make More Profit

Today's marketplace is characterized by global rivalry, cut-throat competition, rapid breakthroughs in product technology, wide fluctuations in raw material costs, and design-to-price strategies. Plus, the customer value and total cost elements of this market environment are often at loggerheads. The goal can no longer be simply to introduce and sell more products—but also to make more profit.

This paradox cannot be resolved by stemming the new product flow. This not only does not solve the contradiction but could even sink the enterprise. The only solution is to understand and eliminate the true causes of cost and maximize the true value for the customer. That is the purpose of VEP—to systematically address long-hidden layers of corporate complexity by finding and dismantling them at their source.

Like a disease, complexity enters the company system much further upstream than anyone ever before suspected. It enters at the point of product conception and infects one process after another as it moves downstream through the organization on its way to the end-user. No one is safe. No department escapes. No company is exempt.

The real goal of VEP—and of this book, *Smart Simple Design/Reloaded*—is to create a shared understanding of negative variety and its solutions. VEP is too important an enterprise recovery process to overlook. Ineffective variety is no longer inevitable when you have learned to systematically avoid the negative aftermath of new product introduction *before* it takes root. Equally, if you want to control—and then reduce, from the inside—your existing parts inventory and the complexity that is its inevitable accomplice, VEP can do that. And if you deploy VEP *while* you initiate new practices and policies, you will prevent negative variety from ever recurring. And that will position your company to grab a genuine and long-lasting competitive advantage—because you will be truly designing products for the bottom line.

Resources

Smart Simple Design/Reloaded

Glossary

Activity-Based Costing (ABC): A system of cost accounting that aligns closely with the objectives and activities of lean production, providing information about total cost and the actual consumption of company resources linked to so-called product "cost-drivers" (e.g., number of purchase orders, customer orders, engineering change notices (ECNs), material moves, machine setups, tools issued to shop floor, product insertions, manual soldering tasks, products shipped); can provide a picture of product costs radically different from data generated by GAAP/traditional accounting systems.

Attribute: Any property, quality, value or characteristic of a part, service or activity.

BOM (Bill of Material): A list of the constituent parts of a product or assembly.

CAD/CAM (computer-aided-design/computer-aided manufacturing): A software-based system that integrates computers into the entire production cycle, from design to fabrication.

Cell Design: A system of shop floor layout that closely groups a series of different machines or operations in sequence into a center or cell, collapsing travel distance and travel time and enabling closely related operational tasks to be completed in sequence and with minimal movement of materials, people or tooling.

CIM (Computer Integrated Manufacturing): A software-driven manufacturing system in which all processes are integrated and controlled by computer, enabling all personnel—product designers and engineers, planners, schedulers, shop floor foremen, and accounting—connected with the process to use the same data.

Control Costs (C-Costs): One of VEP's three categories of True Cost; refers to costs associated with task and information transactions that support the other two VEP cost categories—Variety Costs and Function Costs; C-Costs include costs associated with drafting, ordering, buying, inspecting, transporting, storing, and machine and facility maintenance; roughly equivalent to indirect costs or overhead in scope and variety.

Control Point: Any transaction—paper, electronic or otherwise—which supports the design, procurement, sorting, retrieval, handling, production, assembly or inspection of parts or service architecture (including both nonproduction and production functions).

Deep-Dive Approach: In VEP, a comprehensive, company-wide approach to achieving variety effectiveness in an organization, supported by a full complement of VEP Teams, with immediate and measurable results within 30 to 90 days, and more widespread improvements over a period of 6 to 18 months; compare with the *Discrete Approach.*

DFM (Design for Manufacturability): A software-based system that assists companies so that they can design, manufacture, and assemble their products in the least time and at the least cost.

Discrete Approach: In VEP, a tactical (or limited) implementation approach to achieving variety effectiveness; typically involves a task force of experts who analyze single product lines for reduction; usually requires two to six months to complete; compare with the *Deep-Dive Approach.*

Disinflation: A macro-economic phenomenon that is characterized by rising customer demand and falling product prices; the opposite of inflation (rising consumer demand and rising prices); to combat disinflation, companies launch aggressive cost-cutting because they can no longer achieve profit margins by raising prices.

Ease of Assembly (VAT-4): One of VEP's six analytical techniques; seeks to ensure that all constituent parts that have been upgraded through the other five VATs (VEP analytical tools) are also easy to assemble.

Economic Order Quantity (EOQ): A type of fixed order quantity used when making or purchasing items; reflects the minimum amount of items that need to be produced to absorb acquisition and carrying costs.

Effective Variety (also Variety Effectiveness): The extent to which the variety represented in new products builds profits. Effective variety is the balance point between customer-driven (positive) variety and internally-triggered (negative) variety; as the company rids itself of the negative causes of variety, this point of balance shifts increasingly in favor of the positive. Effective variety means achieving customer-driven variety at the *least* cost.

Fail-Safe/Poka-Yoke Device: A physical device (mechanical, electronic or otherwise) used to 100 percent inspect a part or product at, or near, the source of its fabrication and/or assembly in order to reduce or eliminate the possibility of a defect or an error that may lead to a defect.

Function Costs (F-Costs): One of VEP's three cost categories; refers to costs that are generated as the company furnishes a product with its required functions through parts specifications, values, and dimensions. F-Costs are triggered in the product development process.

GAAP (Generally Accepted Accounting Principles): The traditional approach to cost accounting, codified in the 1930s and still widely in use today; tends to mask complexity and ballooning parts inventories.

GAAP Definition of Profit: The difference between product price and product cost.

GAAP Definition of Product Cost: Labor Costs + Material Costs + Overhead Costs: L+M+O

Industrial Designer (a.k.a. Stylist): Person responsible for conceptualizing and pre-inventing all aspects of a new product—its function, geometry, style, and visual impact; primary focus is realizing the needs of the customer in product form.

Ineffective Variety: The condition that arises in an organization when negative variety outweighs positive variety; signifies that efforts to rid the company of negative (internally-triggered) variety can be strengthened; even when strengthened, some degree of ineffective variety may continue to exist temporarily due to momentary technological or budgetary constraints.

Internally-Triggered Variety: Negative variety; variety that results from policies, practices or requirements internal to the organization and *not* from the customer; the opposite of customer-driven (positive) variety.

Kanban: A visual pull system for parts usage, used to create and ensure minimum levels of WIP and inventory; a supporting methodology, used in conjunction with JIT, cells, and lean production.

Labor Cost: A GAAP term referring to company expenditures incurred in providing the manpower, mechanization, and automation required to convert a parts list (BOM) into the desired level of product; in GAAP, a direct cost.

Least Cost Sum: The total cost figure that represents the least amount of company resources needed to achieve a specific outcome, such as a new product.

Material Cost: A GAAP term referring to the purchase price of each component part or raw material that a product requires, as listed on the Bill of Material (BOM) of that product; in GAAP, a direct cost.

Modularity (VAT-2): One of VEP's six analytical techniques; seeks to combine multiple parts into single units or sub-assemblies that are interchangeable within and across product lines, thereby decreasing total cost and increasing the possibility of creating new products by using already existing units or modules.

MRP (Material Resource Planning): A software-driven method for planning, scheduling, and forecasting all the resources needed in the manufacturing process.

Multi-Functionality and Synthesis (VAT-3): One of VEP's six analytical techniques; *multi-functionality* seeks robust design, formulating product structures that include only the minimum number of parts needed to fulfill required functions, with each part serving as wide a range of functions as possible; *synthesis* seeks to extend product robustness by using new materials, production technologies or structural concepts that allow previously separated parts to be merged, collapsed or eliminated.

Negative Variety: Internally-triggered variety; any variation that is not in direct response to a customer demand, request or interest that is not customer-driven; adds cost, not value; results in escalating organizational complexity and parts inventories.

New England Farmhouse Effect: A VEP analogy to express the similarity between the unplanned and haphazard expansion of a company's product offerings and the way a farmhouse tends to spread—in any shape and direction, as needed at the time.

Non-Value-Adding Activity (NVA): Waste; usually divided into seven categories (Seven Deadly Wastes) of non-value-adding activities: 1) making defects, 2) overproducing, 3) over-processing, 4) material handling, 5) motion, 6) delays, and 7) making inventory. Also includes opportunity loss.

Overhead Cost: A GAAP term referring to a group of expenditures that are allocated across all products on a formulaic basis; an indirect cost; includes depreciation on equipment, heat, light, power, taxes, research and development costs, maintenance costs as well as salaries and wages for operations-support personnel and fringe benefits for direct labor and operations-support personnel.

Part: One of the discrete, constituent elements into which a product can be physically separated.

Parts Attribute Template: In VEP, a grid for capturing the characteristics (attributes) of a part which engineers have identified as primary and necessary for making sound VEP-based decisions when developing new products; such templates are developed for each part in the company's part universe; used by the VEP Parts Type Analysis Team in developing a VEP-capable parts classification system.

Range (VAT-5): One of VEP's six analytical techniques; a statistical measure of dispersion (spread) of a group of values that share some common observable characteristic that shows the difference between the largest and smallest of those values and includes all the values along the way; used in VEP to identify and minimize redundancies and overlaps (or near overlaps) in variations in parts, products, product lines, and processes.

6-VATS: The six VEP analytical techniques (VATs) used by VEP analysis teams: VAT-l/Unique vs. Shared; VAT-2/Modularity; VAT-3/Multi-functionality and Synthesis; VAT-4/Ease of Assembly; VAT-5/Range; and VAT-6/ Trend (see individual listings for definitions).

Statistical Process Control (SPC): A method for gauging the likelihood that the output from a process will fall within acceptable limits; well-known SPC tools include check sheet, cause-and-effect diagram, bar graphs, scatter diagram, and other assorted graphing techniques.

3-View Analysis: VEP's central analytical process for streamlining the company's product line, simplifying product structures, and minimizing part counts; analysis work is divided between three VEP teams: View 1/Market Analysis, View 2/Product Structure Analysis, and View 3/Parts Type Analysis.

Total Cost (according to VEP): Variety Costs + Function Costs + Control Costs.

Trend (VAT-6): One of VEP's six analytical techniques; seeks to identify the pattern and direction of a group of values that share some or several characteristics, laid out in some pre-set order (e.g., ascending value, descending value, etc.); used in ascertaining design, usage, customer, and/or other biases or tendencies in the product development arena.

Tri-Cost/True-Cost Model: VEP's perspective on product and total cost that differentiates between costs incurred by the company: 1) to achieve product function (Functional Costs or F-Costs), 2) to offer/make multiple products (Variety Costs or V-Costs), and 3) to support products, parts, and production and non-production processes (Control Costs or C-Costs).

True Cost: A VEP reference model that differentiates three categories of cost: Function Costs (F-Costs), Variety Costs (V-Costs), and Control Costs (C-Costs) in order to: 1) clarify the levels of organizational complexity brought about by negative variety, and 2) identify opportunities to reduce or dismantle that complexity; aligns more closely to the authentic causes of large part inventories than more traditional cost models found in GAAP; does not (and does not attempt to) provide an exact dollar reckoning of each cost per product.

Unique vs. Shared (VAT-1): One of VEP's six analytical techniques; seeks to identify those parts which are dedicated or unique (aka, variable) and those which are shared or commonized (aka, standardized, carried-over); seeks to minimize the number of unique parts and increase the number of shared or standardized parts.

Value-Adding Activity (VA): Activities that change/transform/convert company resources into a product or service—into something of value that the customer is willing to pay for.

Variety Costs (V-Costs): One of VEP's three cost categories; refers to costs that are triggered when a product line expands or diversifies, even if only one attribute of a single part changes; V-Costs can be customer-driven or internally-triggered.

Variety Effectiveness (also Effective Variety): The extent to which the variety represented in new products builds profits; effective variety is the balance point between customer-driven (positive) variety and internally-triggered (negative) variety; as the company rids itself of the negative causes of variety, this point of balance shifts increasingly in favor of the positive; variety effectiveness means customer-driven variety at the *least cost*.

VEP-Capable Classification System: A company database that houses information on parts and products that is accurate, relevant, complete, and accessible and that supports design and marketing decision making based on VEP goals and principles.

VEP: Variety Effectiveness Process®: A systematic, team-based methodology directed at maintaining or increasing customer selection while reducing negative variety in parts, processes, and control points and preventing its recurrence. VEP's goal is to lower costs dramatically and de-complicate systems while maximizing a company's ability to respond to the demands of the market.

VEP Parts Index: A matrix of information that delineates: 1) all part numbers by parts type in the associated bills of material for a given model or product population, 2) the number of times each part number occurs across that population, and 3) the result of *multiplying* the total number of occurrences *times* the total number of parts in that population. This index model is based on a definition of variety as the *total quantity of parts handled*.

Work-in-Process (WIP): All the materials, parts, and subassemblies that exist on the plant floor between the release of raw material and finished-goods inventory.

X-Type Company: A company that practices effective variety, as evidenced by steadily rising sales, coupled with parallel decreases in costs and part counts.

Y-Type Company: A company that does not practice variety effectiveness, as evidenced by escalating part counts and part inventories—even in the face of increasing sales.

Smart Simple Design/Reloaded

260

GLOSSARY

Visual Thinking Inc. & The Visual-Lean® Institute
Resource Page

Under the leadership of Gwendolyn Galsworth, Visual Thinking Inc. (VTI, formerly Quality Methods International Inc.) is the premier resource around the world for visual workplace principles, practices, products, and services.

Visual Thinking's educational arm, The Visual-Lean® Institute, is dedicated to training and certifying in-house instructors and external consultants in *Smart Simple Design/Reloaded* and any of the Institute's nine core visual workplace methodologies—from operator-led visuality to visual leadership for executives. We also offer a train-the-trainer process for all of our courses, both for the general public and exclusively for your company, on-site.

The Visual-Lean® Enterprise Press is our publishing arm, specializing in books about workplace visuality, strategic improvement, variety effectiveness, and other topics that define and support operational and design excellence. Amazon is our main fulfillment partner, offering our books in print, on Kindle, and globally as Print-On-Demand.

Based on more than thirty years of Galsworth's research in the field, our methodologies are robust, complete, and well-tested. Our clients represent a wide range of industries and settings—from discrete manufacturing to continuous process flow factories, from military depots to government agencies and banks, from the healthcare industry and universities to open-pit mines.

VTI sets the pace for the industry, with offerings across five continents that include seminars, workshops, conferences, on-line/on-demand courses, keynotes, study missions, and complete company conversions—all focused on workplace visuality and smart simple design. Our clients include many Fortune 100 and 500 companies as well as small and medium-sized companies across the supply chain.

With over 70,000 devices in our solutions database, we are continually refining and upgrading our instructional materials, implementation designs, and behind-the-scenes support. We offer a full range of products and services—including over 60 on-demand webinars covering all our courses, DVDs, and other off-the-shelf training packages and training aids. Keynote and conference sessions delivered by Dr. Galsworth—as well as her on-site services—are scheduled through our offices. Contact us directly or visit our website for more.

info@visualworkplace.com
503-233-1784
www.visualworkplace.com

Index

3D printing 5, 112
   see also: additive manufacturing, rapid prototyping
3-View Analysis 44, 128, 131-135, 146, 165, 186-214, 224-232
   definition of 187
30-character alpha-numeric description format 152, 156
95:5 Ratio 33-34
   see also: VA/NVA, value adding

A

activity-based costing (ABC) 57-58, 69, 75-76, 245
active parts count 20, 50
additive manufacturing 5, 59
adjustments, eliminate or simplify 177
aggregate data 181
all costs adhere to the part 61, 69, 128
   see also: one new part, parts trigger costs, single new part
Amazon 93
amplification 172
APEX Product Development Excellence Award 245
Apple 7, 92-93, 111, 114
ashtrays 79, 209
assessment criteria 144-145
A. T. Kearney analysis 244
attribute
   characteristics 99
   template 161-162, 207, 219
average
   is good enough 114
   will sell 116

B

Bain & Company 127
balance point 13, 76, 81-82, 122, 189
ballpoint pen example 62-66
banks 94
bill of material/BOM 63, 65-67, 198, 200-201, 203
   physical visual layout of BOMs 199
Brunner, Bob 111
Borden (Foods) 80, 183
boundary of variation 179
Brisch, E. G. & Partners 158
Business Week 115
by-products 31-35, 37, 44

C

CAD (computer-aided design) 28, 96, 99-100, 107, 112-113, 119, 223
CAD/CAM 96, 223
CADIS Inc. 29
Calsonic 79
cannibalize 90-91, 105
capital equipment 101
car model variations 79

C-Costs (control costs 51, 70-71, 73, 86-87, 97, 215
chronic busyness 20-21
Chuck Wagon 139
Chrysler 114-115
CIM (computer-integrated manufacturing) 112
Clark and Fujimoto study 27
class code 152-153, 155, 163
classification systems 2, 14, 38, 44, 131, 133, 137, 149-165, 179, 187-188, 191, 206, 214, 216, 224, 230-231, 237, 256, 259, 260
Coca Cola 92
commonization 167, 170, 182, 200
competitive advantage 28, 39, 46, 59, 112, 168, 239, 249
complexity 30, 35-37, 41-43, 45
  cost of 14
  measuring 62–68
comprehensive view of cause 42-46
concurrent engineering 3, 28
constituent part 175, 203
consumer 28-29, 58, 80, 89, 183
consumer demand 9-10, 13, 27
consumerism 25
continuous improvement 3, 20-21, 23
control points 31-46, 66-75, 82, 95, 101, 129-143, 166-174, 212-247
  definition of 213
control points reduction team 131-140, 188, 214, 223, 227, 230-232, 235, 252
core products 18
cost
  accounting 53
  constraints 90, 96, 112, 122
  curve 36
  cost-cutting 10, 43
  drivers 75, 121
  indicator 223
  of complexity 14
  -per-piece 85-86, 97-98, 101, 103-104, 106, 108
cross-functional analysis 39

cultural push back 135
Cummins Engine 77
customer
  choice 27
  -driven 13, 18-19, 26, 41
  requests 83, 87-88, 104
  selection 12, 13, 28, 37, 43
cycle-time reduction 20

D

dashboard meters 79, 80, 172, 209
dedicated parts 66, 128, 147, 168
deep-dive approach 45, 127-129, 229
description field 156
design
  coherence 95
  for manufacturability (DFM) 174
  from scratch 96, 99
  from the outside in 114
designers 14, 60, 91, 95, 96, 98, 100, 106-108, 111-117, 120-122, 150
designing products for the bottom line 249
DFMA 248
direct cost 53, 245
direct labor 50, 51-52, 54, 85- 87, 104
discounts 89-90, 105
discrete approach 45, 127-128, 148
dis-economies of scale 86
disinflation 9, 10, 58
disposable diapers 8, 27, 94

E

Early Victories Team 131-133, 138-141, 188, 214, 230, 232, 237
ease-of-assembly 178
ECN 39
economic order quantity (EOQ) 90
economies of scale 28
Education and Methods Team 131, 133, 138, 140, 142, 188, 214, 230

Index

effective variety 13, 15, 38, 42, 44, 49, 76, 81-84, 103, 113, 119-122, 135, 159, 189, 191, 234
E. G. Brisch & Partners 158
engineering-based tools, 42
engineering costs 29
EOQ 90, 100-101, 105, 108
erector sets 173
ERP 101, 153, 157
Escort (Ford) 115
expediters 19

F

F-Costs (function costs) 70-71, 73, 174, 213
flankers 18, 27, 94
flow 21, 23, 34, 37, 38
Ford 115
    Motor Company 7, 26
    assembly line 26
    see also: Escort

forms 140, 158, 223-227, 235
Forms Index 227
Froggie 213
Fujimoto (and Clark) study 27

G

GAAP 14, 49, 52-58, 69, 75, 253
    flaws in 57
gatekeeping functions 227
Boothroyd, Geoffrey and Dewhurst, Peter Product Design for Assembly 165
Boothroyd, Geoffrey and Dewhurst, Peter Product Design for Manufacture and Assembly 177–178
geometric functionality 112
Gillette 78, 111
global 27, 55, 58, 77-78, 118, 245-246, 249
GM 115, 172
Google 93
Group Technology (GT) 158

Grove, Andrew 17
growth strategy 92, 94

H

Hewlett-Packard 245-246
    HP Loveland (Colorado) 245
Hicks, Juanita 150
high-profile commitment 236
histogram 181
holistically 17, 248
Honda 116-117
Hyde, W. F. 158

I

iceberg effect 11, 20
improved policies 84-108, 135
improvement ideas, 203
indirect costs 53, 75, 76
industrial designers 111
ineffective variety 4, 34, 88, 98, 105-106, 108, 244, 254
information-capable parts data system 151
    see also Parts
innovation 10, 77, 93-94, 112, 121-221
interchangeability 171
internally triggered 13, 72, 254
internal vs. external value 121
inventory
    hidden costs 51
    turns 24, 87, 91-92, 98, 101, 104, 106-107, 109, 246
IT 10, 13, 20, 31, 44, 81, 83, 99-100, 107, 135, 139, 141, 143, 152, 155, 157, 164, 189
IT applications 189

J

Japan 23, 59, 79, 116-117, 245
JIT 33
Just Do It! 144

K

Kearney, A. T. analysis 244

Kohdate, Akira and Suzue, Toshiro, Variety Reduction Program 62, 69, 165

L

lean 3-4, 18-19, 21, 25, 33
least cost sum 4, 12, 28, 46, 98, 105, 106, 109
least-sum means 55, 81
Legos 173
Lail, Eric 149-151, 229
life-cycle continuum 145
L+M+O 86-87, 89-90, 96, 98, 104-106, 109, 145
lot sizing 100-101, 108
Lotus cars 115
low intrinsic value 175

M

make vs. buy 85-86, 104
market
    analysis 187, 190-191, 194, 197, 198
    analysis team 131-142, 166, 184, 188-197, 200, 209, 214-215, 230, 237
    attribute matrix 195, 197
    dominance 78
    segmentation 26
    value 77, 89
market-in design 18, 25, 113
marketing 12-13, 15, 19, 29, 44, 46
mass customization 26-27, 80
Matsushita 79
Maxima 245
McTrucks 118
"me-too" products 91, 94
Milacron Inc. 117
Model T, the ultimate universal product 26
Modularity 165, 170-173, 186, 200, 208, 225
module-mating 171
Mohan, Venkat 29
more revenues, less profit 21
Mr. Potato Head 172
multi-functionality 174, 176

N

negative variety 4, 13-15, 37-50, 77-111, 114, 117, 166-167, 182-189, 197, 203, 212-216, 229, 231, 235, 249
    definition of 80
Neon 114-115
new products 8, 10, 12-13, 18, 20, 25, 27-29, 31, 33, 35-37, 41-42, 77
Nike 7, 111, 114, 255
Nissan 79, 80, 117, 167, 209, 244-245, 246
nomenclature 44, 99, 136, 138, 159, 189, 191-192, 209, 219
    product heirarchy 191
    standardizing 138
non-standard parts 168
non-value-adding activity (NVA) 21-24
numerical scoring scale 146
Nypro 246

O

obsolescence 88-89, 105, 128
OEMs (original equipment manufacturers) 79, 184
one new part 25, 29-31, 59, 60-62, 221, 255
    see also: all costs adhere to the part, new part, single new part
operational excellence 18, 25
organizational congestion 35
organizational gridlock 244
overall cost 114, 121
overdesign 121
overhead 11, 49-57, 73, 75, 85- 90, 98, 104-105, 107
overlaps 180, 182, 187, 194, 197, 232

P

paperwork 12, 30, 51, 57, 63, 88, 143, 215, 221-222, 247
part-life-cycle 30
part number prefixes 153

parts classification system 38, 130-31, 133, 137, 151- 164, 188, 191, 206, 214, 230, 231, 237
  see also: classification system
parts count 20, 38, 46
parts index
  see also: VEP Parts Index
parts inventories 10, 12, 46
parts-mating 171
parts trigger cost 134
parts type analysis team 133, 135-136, 162, 166, 206, 208-209, 237
  definition of 204
part symmetry 177
pattern of variation 179
performance measures 54
Pine, Joseph, Mass Customization 26, 80
platform team 115
poka-yoke 23, 100, 238
policies 4, 5, 13, 18, 37, 42-46, 77-109, 113, 120, 123, 127, 129, 134-135, 185, 224, 248, 249
  formal and informal 14, 39, 43, 81, 83, 103, 134, 220, 231

policy team 133-135, 143, 231, 237
political sensitivity 134
positive variety 13, 45, 80, 81
  definition of 80
prefix rule 156
prevention-based mindset 42
process
  definition of 213
process analysis team 215, 219, 220
process attribute template 219
process reduction team 131, 133, 136-137, 188, 214, 230, 231
Proctor & Gamble 92
product-attribute matrix 194
product
  definition of 198
  analysis procedure 198
  complexity 59, 66, 121, 173, 198
  design function 120

diversification 26, 39, 58, 72, 74, 92, 189-190, 244
engineers 111, 161
expansion 18, 21, 32-33, 37, 40, 45
fragmentation 95
nomenclature 44
obsolescence policy 88-89, 105
proliferation 28, 69, 78, 80, 151
series 50, 62, 128, 132, 144-148, 168, 170, 189, 191-197, 215, 245
simplification 203
structure 46
universe 68, 96, 118, 142, 146, 194, 197-198, 203, 247
variety 10, 27, 36, 43, 66, 76, 78-79, 81, 179, 182
product structure analysis team 131, 135-137, 166, 188, 199, 208, 214, 230, 237
production process 12
product life cycles 8, 28, 61, 88, 97, 105, 113
  collapsing 28
profit erosion 36, 244
profit-making 35
profit margin 24, 58, 90-91, 105, 117
PUI case study
  kickoff 132
  1. more revenue, less profit 17–19
  2. negative variety 36–37
  3. accounting issues 50–51
  4. parts index 66–68
  5. Sid, sharing parts 119-120
  6. selecting target series 146, 146–147
  7. classification system 151–156
  8. market analysis 191-197
  9. product analysis 198-203
  10. parts type analysis 206-209
  11. production processes 215-220
  12. control points 222–223
PUI
  parts type team 163

product structure analysis team 199
market analysis team 193, 197
parts and product classification system 152
parts type analysis team 206
product levels 192
steering team 232
top 20 192-194, 197

Q

QFD 96-97, 113, 245
qualifying process 144-148

R

radiators 79, 209
range
  definition of 179
rank order 144
Rapid Prototyping 112
rate-of-parts increase 35
rate-of-revenue increase 35
ratio of shared over unique 168
R&D departments 78
reactive approach 95
ready, fire, aim 144
reduction proposals 130-131, 134, 136, 142, 188, 196, 203, 214, 219, 227, 230
redundancies 180, 187, 194, 197, 232
relations diagram 145
replication 172-173
rocks,
  in the river of flow 23
routing sequences 158
runaway by-products 34-35, 37, 44

S

Saturn 115
Schonberger, Richard 187
scope 73, 127, 129, 136, 151, 158, 165, 170, 174, 229, 231
secondary level of cause 25
select your starting point 144
self-aligning 177

self-locating 177
selling price 9, 58, 89-90, 105, 117
service offerings 189
seven deadly wastes 22-23, 33-34
shared (parts) 63, 66, 72, 106, 109, 114, 116, 119, 121, 135, 138, 145, 166-168, 170, 172-173, 194, 200, 207, 219, 249
Sid (designer at PUI) 119
single part 4
single new part 28-30
  see also: all costs adhere to, one new part
six sigma 25
six VATs 46, 130-131, 164-165, 185-189, 194, 198-199, 200, 203, 212-214, 219, 224, 230-231, 235
SKU portfolios 128, 189, 244
SMED 158, 238
software 29, 36, 78, 113, 152-153, 158, 232
special fixtures 88, 168
specification information 99
stakeholders 46, 134, 135, 196, 200, 204, 209, 232
standardization 26, 72, 138, 158, 168, 170-171, 173
standardize within a range 181
standard modules 171, 173
status quo 102, 134
steering team 131-134, 140-146, 168, 188-189, 196, 203, 209, 214, 216, 220, 227, 229-232, 234-235, 237
steering team calendar 234
steering wheels 79-80, 136, 204
stress 243
summative scale (rank-order continuum) 146
supplier base selection 84
suppliers 20, 46, 50-51, 55, 66, 79, 84-85, 101, 103, 108
surface development 120
Suzue, Toshiro and Kohdate, Akira, *Variety Reduction Program* 62, 69, 165
SWAT team 118

synthesis 174, 176

T

targeted line 144
target-pricing 117
team-based approach 38, 42-43
technology breakthroughs 77
terminology 152, 159, 191, 193, 216, 224
Tesla Motors 111
time-to-market 28, 113
   collapsing 28
top-down assembly 177
total active parts 38
total cost 29, 52, 55-57, 72, 74, 86, 90, 105, 121, 249
Toyota 23
TPM 25
Traditional
   accounting procedures 44
   allocation systems 87
   definition of profit 56
transaction 73, 137, 220
trend 165, 179, 183-186, 209, 225
Tri-Cost Model 14, 49, 62, 69, 75-76
   see also: VEP Tri-Cost/True Cost Model
triggers of negative variety 83–110
true cost 49, 55, 58–75, 64, 69, 70, 89, 96-98, 106, 108, 145, 223, 232, 246
TRW Inc. 115

U

unified strategy 120, 122
unifying development strategy 44, 92, 179
unique/dedicated parts 63, 66, 79, 106, 109, 116-117, 121, 147, 166-168, 170, 190, 203, 219
unique vs. shared-parts ratio, 121
United Electric Controls Company 246
unrestricted vision 177
unwarranted parts 34

V

value-adding activity (VA) 21-24, 33
   VA/NVA ratio 23
value engineering 43, 71, 85, 96-98, 106-108, 247
value engineering/value analysis (VE/VA) 71
Vanderstoop, Nick 247
variety
root cause of 59
V-Costs (variety costs) 71-72, 74, 174
variety effectiveness 12-13, 37- 46
   definition of 40-41
VAT-1 165-168, 170-172, 177, 185, 190, 194, 200, 207
VAT-2: Modularity 165, 170-173, 177, 186, 200, 208
VAT-3: Multi-functionality & Synthesis 165, 174-177, 186, 194, 201, 208
VAT-4: Ease of Assembly 165, 177-179, 186
VAT-5: Range 165, 179-183, 186, 207
VAT-6: Trend 165, 179, 183-186, 209
Venerable Chair Company
   case study 149–151
Venmar 191
VEP analysis teams 134
VEP Awareness Sessions 133, 140, 229, 237
VEP-capable 14, 148, 151, 159, 179, 187, 206, 231, 237
VEP champion 134
VEP Methodology 14-15, 37-46, 127-131, 142, 164, 188, 213-214, 229- 237, 246
   definition of 37
   overview of 129–130
VEP mindset 119, 131, 149, 188, 214, 230, 239, 244
VEP Parts Index 49, 62–69, 76, 140, 145, 161, 190, 196,

199, 201-202, 206-207, 224
  as relative measure of complexity 63–64
  as improvement driver 69
VEP proposal impact worksheet 232-233
VEP review calendar 235
VEP, rewards of 38–39
VEP team fact sheet 141–144
VEP teams 14
  overview 133–142
VEP Tri-Cost/True-Cost Model 49
  see also: Tri-Cost Model
verifiably customer-driven 97, 166
visual 3-4,
visuality 3, 18, 25
visual layouts 203

W

waste 3, 19-24, 31-32, 43

waste elimination 20-21
watchdog 135, 235
weeding out duplicates, near-duplicates, and other extraneous parts 158
weighting formula 144
when average is good enough 114
word-of-mouth promotion 139

X

X-type 41-42
X-type profile 41-42

Y

Y-shaped curve 35, 41
Y-type trajectory 36

Z

zero defects 25

About Gwendolyn D. Galsworth

Gwendolyn Galsworth, PhD, is president/founder of Visual Thinking Inc. (VTI, formerly Quality Methods International Inc.), a training, research, and consulting firm—and of The Visual-Lean Institute®, which offers licensing and train-the-trainer in all of its courses, including Smart Simple Design. The Visual-Lean Enterprise Press is the company's publishing arm.

Across over 30 years of hands-on implementations and research, Dr. Galsworth developed, codified, and refined the technologies of the visual workplace. The result is a coherent system of logic and application that contributes powerfully and distinctly to enterprise excellence, a continuous improvement work culture, and trackable bottom-line results. The original *Smart Simple Design (1994)* was written in parallel to her work in workplace visuality. *Smart Simple Design/Reloaded (2014)* is the result of her exploration of the field of strategic improvement since then.

Her work has been honed and tested in companies around the world—factories, banks, hospitals, military depots, and offices—including Lockheed-Martin/USA, Pratt & Whitney/USA, Johnson & Johnson Surgical Instruments/Mexico/USA, Royal Nooteboom Trailers/Holland, Moog Quick Set/USA, Trailmobile/Canada, Beth-Israel Deaconess Medical Center/USA, Flinders Medical Group/Australia, Rolls-Royce/UK, Crompton Greaves/India, Sears Roebuck/USA, Wilson Transformer/Australia, and United Electric Controls/USA.

It was her 1993 collaboration with David Reis and Bruce Hamilton at United Electric Controls that resulted in the original *SSD* book—and the new *Smart Simple Design/Reloaded*.

Among her many other landmark works, Galsworth is the author of *Visual Workplace/Visual Thinking* and *Work That Makes Sense*, both winners of the Shingo Research Prize. She is also the author of a wide range of training systems and materials in many media—audio, video, and online.

Her on-demand webinars and off-the-shelf training packages are rich in content and insight, each containing scores of actual solutions and a step-by-step process for developing more.

Hear Dr. Galsworth every week on her popular radio show, *The Visual Workplace*, on www.VoiceAmerica.com. Podcasts are available at www.visualworkplace.com.

info@visualworkplace.com
503-233-1784
www.visualworkplace.com

Made in the USA
San Bernardino, CA
09 November 2014